The Environment and Sustainable Development in the New Central Europe

AUSTRIAN AND HABSBURG STUDIES
General Editor: Gary B. Cohen, Center for Austrian Studies,
University of Minnesota

Volume 1
*Austrian Women in the Nineteenth and Twentieth Centuries:
Cross-Disciplinary Perspectives*
Edited by David F. Good, Margarete Grandner, and Mary Jo Maynes

Volume 2
*From World War to Waldheim: Culture and Politics in Austria and
the United States*
Edited by David F. Good and Ruth Wodak

Volume 3
Rethinking Vienna 1900
Edited by Steven Beller

Volume 4
*The Great Tradition and Its Legacy: The Evolution of Dramatic and Musical
Theater in Austria and Central Europe*
Edited by Michael Cherlin, Halina Filipowicz, and Richard L. Rudolph

Volume 5
*Creating the "Other": Ethnic Conflict and Nationalism in Habsburg
Central Europe*
Edited by Nancy M. Wingfield

Volume 6
Constructing Nationalities in East Central Europe
Edited by Pieter M. Judson and Marsha L. Rozenblit

Volume 7
The Environment and Sustainable Development in the New Central Europe
Edited by Zbigniew Bochniarz and Gary B. Cohen

THE ENVIRONMENT AND SUSTAINABLE DEVELOPMENT IN THE NEW CENTRAL EUROPE

Edited by

Zbigniew Bochniarz

and

Gary B. Cohen

Berghahn Books
NEW YORK · OXFORD

First published in 2006 by
Berghahn Books

www.berghahnbooks.com

© 2006 Zbigniew Bochniarz and Gary B. Cohen

All rights reserved.
Except for the quotation of short passages
for the purposes of criticism and review, no part of this book
may be reproduced in any form or by any means, electronic or
mechanical, including photocopying, recording, or any information
storage and retrieval system now known or to be invented,
without written permission of the publisher.

Library of Congress Cataloging-in-Publication Data

The environment and sustainable development in the new Central Europe / edited by Zbigniew Bochniarz and Gary B. Cohen.
   p. cm. — (Austrian and Habsburg studies ; 7)
  "The essays in this volume represent revised and expanded versions of papers selected from a conference organized by the Center for Nations in Transition at the University of Minnesota, Twin Cities, September 19–21, 2003"—P. xii.
  Includes bibliographical references and index.
  ISBN 1-84545-144-9 (alk. paper)
   1. Sustainable development—Europe, Central—Congresses. 2. Europe, Central—Environmental policy—Congresses. I. Bochniarrz, Zbigniew. II. Cohen, Gary B., 1948– III. Series.

HC244.Z9E519 2006
338.943'07—dc22

                                                        2006040780

British Library Cataloguing in Publication Data

A catalogue record for this book is available from the British Library

Printed in the United States on acid-free paper

ISBN 1-84545-144-9 hardback

Contents

List of Figures and Tables	vii
Acknowledgements	xi
Notes on Contributors	xiii
Introduction: Legacies, Challenges, and New Beginnings *Zbigniew Bochniarz and Gary B. Cohen*	1
1. From Communism to Climate Change: The Sustainability Challenge and Lessons from Central Europe *Robert Wilkinson*	11

Part One: The Environment as Policy Priority

2. Assessing Sustainability of the Transition in Central European Countries: A Comparative Analysis *Sandra O. Archibald and Zbigniew Bochniarz*	19
3. Sustainability and EU Accession: Capacity Development and Environmental Reform in Central and Eastern Europe *Stacy D. VanDeveer and JoAnn Carmin*	45
4. Sustainability of Clusters and Regions at Austria's Accession Edge *Edward M. Bergman*	59

Part Two: The Economics of Sustainable Development

5. Greenhouse Gases Emissions Trading in the Czech Republic *Jiřina Jilková and Tomáš Chmelík*	81
6. Ecological Reform in the Tax System in Poland *Olga Kiuila and Jerzy Śleszyński*	99

Part Three: Water Policies and Institutions

7. The Czech Republic: From Environmental Crisis to Sustainability   121
   Václav Mezřický

8. The Tisza/Tisa Transboundary Environmental Disaster:
   An Opportunity for Institutional Learning   129
   Eszter Gulácsy, László Pintér, and Jim Perry

9. Austria and the EU Water Framework Directive   143
   Wilhelm R. Vogel

10. The Western Bug River: UNECE Pilot Project   150
    James B. Dalton, Jr.

11. Wastewater Treatment in the Postcommunist Danube River Basin   167
    Igor Bodík

Part Four: Agriculture and Rural Development

12. "Thinking Unlike a Mountain": Environment, Agriculture, and
    Sustainability in the Carpathians   183
    Anthony J. Amato

13. New Approaches to Sustainable Community Development in
    Rural Slovakia   203
    Slavomíra Mačáková

14. Sustainable Development in Moravia: An Interpretation of the Role
    of the Small-Town Sector in Transitional Socioeconomic Evolution   217
    Antonín Vaishar and Bryn Greer-Wootten

15. Building Local Sustainability in Hungary: Cross-Generational
    Education and Community Participation in the Dörögd Basin   232
    Judit Vásárhelyi

Select Bibliography   247

Index   257

List of Figures and Tables

Chapter 2

Figure 2.1.	GNP per Capita	24
Figure 2.2.	GNP per Capita—Liberalizers	25
Figure 2.3.	GNP per Capita—Financial Reformers	25
Figure 2.4.	GNP per Capita—Category 1 Privatizers	26
Figure 2.5.	GNP per Capita—Category 2 Privatizers	26
Figure 2.6.	Contribution of Services to GDP—Liberalizers	27
Figure 2.7.	Contribution of Services to GDP	27
Figure 2.8.	Contribution of Services as Percent of GDP—Stabilizers	28
Figure 2.9.	Contribution of Services to GDP—Reformers	28
Figure 2.10.	Contribution of Services to GDP—Early Privatizers	29
Figure 2.11.	Contribution of Exports of Goods and Services to GDP	29
Figure 2.12.	Contribution of Imports of Goods and Services to GDP	30
Figure 2.13.	Life Expectancy at Birth	31
Figure 2.14.	Fertility Rate	31
Figure 2.15.	Infant Mortality Rate	32
Figure 2.16.	Carbon Dioxide Emissions	33
Table 2.1.	Average Annual Deforestation 1990–2000	33
Table 2.2.	Factors of Environmental Significance Following Transition	37
Table 2.3.	Key Air Quality Pressure Indicators	40
Table 2.4.	Econometric Results	41
Table 2.5.	Income Levels at Emissions Turning Points	42

Chapter 4

Figure 4.1.	Survey Form for Austrian Firms (2001)	76
Table 4.1.	Sources of Innovative Inputs	69
Table 4.2.	Strong Cluster and Strong Region Advantages	70
Table 4.3.	Orientation of Clusters and Regions toward Competition and Differentiation	71

Chapter 5

Figure 5.1.	Total Primary Energy Supply (TPES/GDP, toe per thousand 1995 USD, exch. rate)	82
Figure 5.2.	Total Primary Energy Supply (TPES/population, toe per capita)	83
Figure 5.3.	CO_2 Emissions from Relevant Sectors 1990–1999 *(Mt CO_2)*	86
Figure 5.4.	Inventories and Updated BAU Projection Scenarios for 2000–2010	86
Table 5.1.	Air Pollution in Selected Countries (1997)	84
Table 5.2.	Emissions and Sinks of CO_2 from Relevant Sectors, 1990–1999 *(Mt CO_2)*	85

Chapter 6

Figure 6.1.	Marginal Costs of Production Inputs	106
Figure 6.2.	Coal Prices *(C)* after Taking into Account Taxes and Costs of Pollution Reduction	107
Figure 6.3.	Marginal Costs of Production	108
Figure 6.4.	Shadow Price for Energy-generating Inputs	108
Figure 6.5.	Shadow Prices of Production Inputs (capital, labor, and energy inputs)	109
Figure 6.6.	Household Demand in Billions PLN (logarithmic scale)	110
Figure 6.7.	Production Volume in Billions PLN	110
Table 6.1.	Environmental Protection Investment in Poland	101
Table 6.2.	Sources of Atmospheric Emissions in Poland in 1995 [%]	103
Table 6.3.	Scenarios Considered and Researched in the CGE Model	105
Table 6.4.	Change in the Number of Employed and Unemployed in Individual Scenarios as Compared to Scenario I (%)	107
Table 6.5.	Total Sulfur Dioxide and Total Carbon Dioxide Emissions Relative to Scenario I [%]	111
Table 6.6.	The Values Reflecting Welfare Changes (measured using CV or EV) for Specific Groups of Households	112
Table 6.7.	Economic Indicators (not taking into account inflation)	113
Table 6.8.	Characteristics of the Model	115
Table 6.9.	Classification of the Economic Sectors in the Model	116

Chapter 7

Table 7.1.	Ecological footprints of the CR and Neighboring Countries	126

Chapter 8

Figure 8.1.	Cyanide Concentrations in the Tisza/Tisa River in the Wake of the Baia Mare Disaster	132

Chapter 9

Figure 9.1.   Administrative Units (Länder) in Austria   146
Figure 9.2.   Basins and Sub-basins in Austria   147
Figure 9.3.   The Danube River Basin   148

Chapter 10

Figure 10.1. Relief Map of Bug River Basin   152
Figure 10.2. Monthly Flow Characteristics for the Western Bug River
              at Wyszkowie from 1951 to 1990   154

Table 10.1.   Summary of Major Water Management Programs for the
              Western Bug River   162

Chapter 11

Figure 11.1. Location of Danube River Basin Countries   168
Figure 11.2. Development of Domestic Water Consumption in the Slovak
              Republic (l/cap.d, 1990–2000)   172
Figure 11.3. Development of Inhabitants' Connection to Sewage Systems
              in the Slovak Republic   175
Figure 11.4. Distribution of WWTPs among Slovak Settlements   176

Table 11.1.   Basic Geographical and Economic Parameters of the
              Danube Countries   169
Table 11.2.   Population and GDP in the DRB Countries.   171
Table 11.3.   Basic Characteristics of the Drinking Water Supply in
              DRB Countries   171
Table 11.4.   Production and Treatment of Wastewater in the
              DRB Countries   174
Table 11.5.   Connection to Sewage System in Selected OECD Countries
              (year 2000)   174
Table 11.6.   Expected Reduction of Pollution from (tons/year) and
              Expected Investment Cost in Municipal WWTPs in the
              DRB for Years 2000–2005   177

Chapter 12

Figure 12.1. A typical Hutsul village in the Carpathians. Photo by Anthony
              Amato and Felixa Amato.   186

Chapter 14

Table 14.1.   Population by Rank-Size of Communities (2001)   220

Acknowledgements

The essays in this volume represent revised and expanded versions of papers selected from a remarkable conference organized by the Center for Austrian Studies and the Center for Nations in Transition at the University of Minnesota, Twin Cities. Some thirty-eight scholars from the United States, Canada, the United Kingdom, Austria, the Czech Republic, Poland, Slovakia, and Hungary contributed papers for the conference, which took place September 19–21, 2002, on the Minneapolis campus of the university. Such a gathering of diverse talents and perspectives was only possible with the generous financial and logistical support of a number of academic and governmental agencies. The Austrian Federal Ministry of Education, Science, and Culture and the Austrian Cultural Forum in New York assumed the travel costs for the presenters who came from Austria and Central Europe. Generous grants from a number of departments and institutes in the University of Minnesota helped defray the local costs: the College of Liberal Arts Scholarly Events Fund; the Consortium on Law and Values in Health, Environment, and the Life Sciences; the Department of Ecology, Evolution, and Behavior; the Department of Economics; the Department of Political Science; the Department of Sociology; the Hubert H. Humphrey Institute of Public Affairs; the International Programs Office of the Carlson School of Management; the International Studies Title VI Grant (Institute for Global Studies): MacArthur Interdisciplinary Program on Global Change, Sustainability, and Justice; the Office of International Programs; the Program in Agricultural, Food, and Environmental Ethics; and the Western European Title VI Grant (European Studies Consortium). The Minnesota Trade Office helped organize a business seminar to accompany the conference and assisted in securing speakers for that seminar and for lunch sessions during the conference. At the Center for Austrian Studies, Barbara Krauss-Christensen and Leo Riegert managed with great aplomb the logistics of holding the conference and transporting and lodging the presenters. Also at the Center, Dan Pinkerton edited announcements and brochures for the conference as well as some of the maps and graphs that appear in this volume.

For the preparation of this volume, Mr. Arnold Lelis of the Center for Austrian Studies deserves special thanks for his efficient, clear-sighted, and patient work in carrying out the challenging tasks of corresponding with authors scattered across Europe and North America and coordinating the editing of the revised essays. In the final stages, Jules P. Gehrke efficiently discharged the work of coordinating the proofreading and corrections. At Berghahn Books, Marion Berghahn and her colleagues displayed the usual high editorial and production standards of the house in the publishing of this volume.

Zbigniew Bochniarz
Gary B. Cohen
Minneapolis, May 2005

Notes on Contributors

Anthony J. Amato is an associate professor at Southwest Minnesota State University's Center for Rural and Regional Studies. He is the author of "The Flow of History: An Essay on Rivers, the Past, and the Present" in *Cloudy-sky Waters: An Annotated Bibliography of the Minnesota River,* by Kris Bronars Cafaro (Marshall, MN: Southwest Minnesota State University, 2004).

Sandra O. Archibald is professor at the Daniel J. Evans School of Public Affairs at the University of Washington in Seattle. Her research interests concern a wide area focusing on the intersection of economics, institutions, and public policy. She is extensively involved in public service and research designed to support sound public policy and management decisions and has worked extensively in Central and Eastern Europe designing environmental curriculum and academic programs for higher education.

Edward M. Bergman has, since 1995, been Professor of Regional Science and Development and Director of the Department of City and Regional Development at the Vienna University of Economics and Business Administration. He has conducted many studies of regional economic development both in Europe and in the United States, and has published widely on these topics.

Zbigniew Bochniarz has been affiliated with the Hubert H. Humphrey Institute of Public Affairs at the University of Minnesota since 1986. In 1994, he founded there the Center for Nations in Transition, which has become an international leader in designing and delivering foreign assistance programs for Central and Eastern Europe in areas of educational, environmental, and institutional reforms for sustainable development. His teaching and research focus on the economic, environmental, and social aspects of sustainability within the transformation processes in post-communist countries. He is the author, co-author, or editor of over one hundred books and other publications on these topics.

Igor Bodík currently teaches and researches in the Department of Environmental Sciences at the Slovak University of Technology in Bratislava. His research focuses on aspects of wastewater treatment technologies and on sustainable technologies. He has numerous publications, including over forty articles in peer-reviewed journals, and has participated at many international conferences.

JoAnn Carmin is currently assistant professor in the Department of Urban Studies and Planning at the Massachusetts Institute of Technology. Her research concerns civil society participation in environmental governance; environmental movements and organizations; environmental policy learning in response to the Czech floods of 2002; and the impact of transnational diffusion on environmental organizations. Among her most recent publications are *Collaborative Environmental Management: What Roles for Government?* with Tomas Koontz, et al. (Washington, D.C.: Resources for the Future, 2004) and *EU Enlargement and the Environment: Institutional Change and Environmental Policy in Central and Eastern Europe,* edited with Stacy D. VanDeveer (London: Routledge, 2005).

Tomáš Chmelík is pursuing his doctoral degree in the Faculty of Economics and Public Administration at the University of Economics in Prague. Since 1998, he has worked for the Ministry of Environment of the Czech Republic, Department of Environmental Economy, Unit of Economic Instruments. His expertise centers on tax reform and emissions trading as instruments for greenhouse gas reduction.

Gary B. Cohen has been director of the Center for Austrian Studies and professor of history at the University of Minnesota, Twin Cities, since 2001. He teaches and publishes on modern Central European social and political history. He is the author of numerous articles and essays as well as two books, *The Politics of Ethnic Survival: Germans in Prague, 1861–1914* (1st ed., Princeton University Press, 1981; rev. 2nd ed., Purdue University Press, 2005) and *Education and Middle-Class Society in Imperial Austria, 1848–1918* (Purdue University Press, 1996).

James B. Dalton, Jr. is Assistant Professor of Geography at the United States Military Academy at West Point. His research focuses on the intersection of geographical and resource management studies with security issues.

Bryn Greer-Wootten is a professor in the Faculty of Environmental Studies and Department of Geography at York University, Toronto, Canada. Her research concerns include the study of elite, scientific, and popular discourse in relation to environmental issues and the ways that environmental policy translates to the local level.

Eszter Gulácsy is Principal Sustainable Technologies Advisor at the Centre for Sustainable Engineering in Peterborough, United Kingdom. Her professional interests include renewable energy sources and EU and national environmental policy; at present, she is working on projects to develop sustainable markets in the countries of Central and Eastern Europe. She has an MS in Environmental Engineering Science from the University of California at Berkeley. She is a coauthor of *International Law and the Baia Mare Cyanide Spill: Final Report*, published by the Regional Environmental Center for Central and Eastern Europe in 2001.

Jiřina Jilková has been Associate Professor in the Department of Environmental Economics at the University of Economics in Prague since 1990 and Executive Director of the Institute of Economic and Environmental Policy there since 1997. She specializes in environmental economics and associated policy issues, and in agricultural economics.

Olga Kiuila is Assistant Professor in the Economics Department at Warsaw University. Her Ph.D. dissertation (Warsaw University, 2000) concerned the *Costs of Adapting Poland's Economy to the II Sulfur Protocol: A General Equilibrium Model.*

Slavomíra Mačáková received her Ph.D. in Environmental Technologies from the Technical University of Kosice, Slovakia, in 1998. Since 1997, she has been a freelance environmental consultant and a researcher and trainer. Particularly, she has worked for ETP Slovakia as program manager and instructor for many projects involving environmental management at local levels and community development.

Václav Mezřický is Associate Professor of environmental law and policy at the Faculty of Law, Charles University, Prague. From 1992 to 1996 he was a member of the Environmental Advisory Committee of the President of the European Bank for Reconstruction and Development. Since 1993 he has been head of the Department of Environmental Law within the Faculty of Law, Charles University, and Deputy Director of the Institute for Environmental Policy in Prague. He has participated in drafting the Act on Environment of the Czech Republic and worked on approximating Czech environmental law with EC legislation (PHARE/DISAE).

Jim Perry is professor and head of the Department of Fisheries, Wildlife, and Conservation Biology at the University of Minnesota. His research interests include ecological approaches to water quality management and policy; water resources education and decision making in developing countries; and environmental management and training in Central and Eastern Europe.

Dr. László Pintér has been with the International Institute for Sustainable Development in Winnipeg, Canada, since 1994, where he is currently Director of the Measurement and Assessment program. His primary areas of interest include the conceptualization, establishment and use of integrated indicator, assessment and reporting systems from the global to the local levels. He is also well known for his work in training and capacity building for integrated environmental assessment and maintains an active interest and project portfolio focused on integrated vulnerability assessment and adaptation to global change.

Jerzy Śleszyński is a professor in the department of Industrial Informatics and Economic Analyses, Faculty of Economic Science, at Warsaw University. His areas of specialization include: environmental economics and environmental protection policy; economic instruments and multiple criteria in the management of natural resources; economic valuation of natural resources; sustainable development policy; and specific single indicators of sustainable economic welfare. He has published widely on these issues and been a consultant on several applied projects both in Poland and abroad.

Antonín Vaishar is a research scientist and Director of the Institute of Geonics at the Brno branch of the Czech Academy of Sciences. His research focuses on population and urban geography, and on environmental problems in regional geographic contexts. He has been the coordinator of several projects, including: Cultural and Economic Conditions of Decision Making for Sustainable Cities; New Prosperity for Rural Regions; Danube Pollution Reduction Program; Floods, Landscape and People in the Morava River Catchment Area; and The Geography of Small Moravian Towns.

Stacy D. VanDeveer is Associate Professor in the Department of Political Science at the University of New Hampshire. His interests include international relations, international organizations, environmental politics, U.S. foreign policy, and European transition politics, especially as these involve issues of sustainability and environmental protection. He has published numerous articles and been co-editor of two volumes on these topics, including most recently *EU Enlargement and the Environment: Institutional Change and Environmental Policy in Central and Eastern Europe,* edited with JoAnn Carmin (London: Routledge, 2005).

Judit Vásárhelyi received her Ph.D. from Eötvös Lorant University, Budapest, in 1974. She has a long record of grassroots and NGO activism and has held official posts as a founding member of the Danube Circle (1984–1989), Soros Foundation program coordinator (1987–1989), member of the Board of Directors at the Regional Environmental Center for Central and Eastern Europe (1990–1993), executive of the Independent Ecological Center (1990–1999), director at the Institute for Sustainable Communities (Vermont) (1995–2001), and member of the Hungarian Committee on Sustainable Development (1996–2001). Since 2001 she has been a board member at Védegylet (Association for Protection), co-President of the Hungarian Association of Environmental Educators, and a freelance lecturer and writer.

Dr. Wilhelm R. Vogel has been at the Austrian Federal Environment Agency since 1989 and has served as Director of the Department of Aquatic Ecology/ Water Protection since 1992. He manages the Austrian Water Quality Monitoring System, investigates human influences on the aquatic ecosystem, and oversees development of indicators in this field (including methods of green accounting), reporting to a wide range of Austrian government offices and ministries. He has applied his experience in national and international water-related information systems and monitoring network development to many EU contexts, including the European Topic Centre on Inland Waters, the Water Framework Directive, and the project "Evaluation of the costs of groundwater inspection in the Member States." Dr. Vogel has published many scientific papers, mostly on environmental pollution and monitoring methods; he is also a lecturer at the University of Vienna.

Dr. Robert Wilkinson is currently a lecturer in the Environmental Studies Program, and the Donald Bren School of Environmental Science and Management, at the University of California, Santa Barbara. He also advises various California and U.S. government agencies on water policy, climate research, and other environmental policy issues; works with the Rocky Mountain Institute, the California Environmental Dialogue, and the Aspen Institute; and consults for corporations, governments, foundations, and nonprofit organizations in the U.S. and internationally. In 1990, Dr. Wilkinson established and directed the Graduate Program in Environmental Science and Policy at the Central European University based in Budapest, Hungary.

INTRODUCTION
Legacies, Challenges, and New Beginnings

Zbigniew Bochniarz and Gary B. Cohen

Just sixteen years ago, the dismantling of the Cold War division of Europe began with parliamentary elections in Poland. Then, like dominos, the other communist Central European countries—Hungary, East Germany, Czechoslovakia, Bulgaria, and Romania—followed, shattering the foundations of the oppressive, totalitarian system and proclaiming democracies and market economies. Sixteen years is but a moment in the long history of Central and Eastern Europe, but for historians of the region the last twelve years of the twentieth century will be remembered as equal in importance to the first years of building independent statehood after World War I. Despite some differences, Czechoslovakia, Hungary, and Poland followed a similar path of development during the last century. Variations were caused by their own internal factors, but outcomes were determined to a great extent by larger geopolitical forces. Today they are jointly entering the twenty-first century with excellent opportunities for future development supported by secure borders with NATO safeguards, democratic political systems, reformed economies, improved environmental quality, and membership in the European Union. Austria, in contrast, had a much different experience as a neutral state during the Cold War, open to the West but trying also to maintain relations with its communist neighbors and even finding some advantage as an almost invisible "middleman" in facilitating confidential hard-currency transactions between communist-bloc members. Since 1989, Austria has supported the transitional processes and mediated these countries' integration into the European community. It has developed closer relations with them in several areas, including environmental concerns in the Danube basin.

Significant environmental challenges accumulated in Central and Eastern Europe (CEE) during the Cold War era. Some of these had roots in nineteenth- and early twentieth-century economic development, but the particularly rapid industrial development strategies pursued by the communist governments in their first two decades created a host of long-term problems. There was massive pollution of water, air, and the soil in many localities. Greenpeace reported in the late 1980s, for instance, that half of the eight hundred some Polish communities along the Vistula River had no sewage treatment facilities. Along more than 80 percent of that river's length, the water was too polluted for even industrial use, and as a result much of the fish in the Baltic Sea, into which the Vistula empties, was unfit for human consumption. The heavy use of soft coal contributed to serious problems with acid rain in many areas and decimated much of the once dense forests in the southern part of the German Democratic Republic, northwestern Czechoslovakia, and Polish Silesia—the "Black Triangle," as the region was called in the international environmental literature. Bitterfeld in the GDR, the region of Katowice in Polish Silesia, and the older Czech industrial centers may have been the pride of communist industry in their respective countries, but the catastrophic environmental pollution in many of those localities caused major health problems for several decades before 1989. Environmental disasters, such as the April 1986 reactor incident in Chernobyl, Ukraine, or the cyanide spill in Romania into the Tisza/Tisa River in early 2000, only added to the environmental problems.

Increasingly during the 1980s the communist governments had to recognize the mounting environmental difficulties and at least promise improvements, but their declining economies could ill afford the costs of remediation and alternative technologies. The limitations of international economic cooperation in communist CEE, whether on a bilateral basis or through the Council of Mutual Economic Assistance (CMEA or Comecon), also made it extremely difficult to take any of the remedial steps on a regional basis that were clearly needed. Unrelieved, the environmental problems became yet another source of popular discontent and found their way into the agenda of dissident groups. Most famously in Bulgaria, for example, environmental complaints fueled the development of a vigorous oppositional group, which began in 1987 and eventually organized on a broader national basis as Ecoglasnost.

Austria enjoyed generally higher environmental standards in the 1980s and early 1990s, but the environmental and general economic conditions of its communist and postcommunist CEE neighbors created significant challenges. During the Cold War, Austria pursued bilateral economic and trade relations with communist neighbors wherever possible, but there were recurring frictions. The seasonal exchange of electrical power between Austria and Czechoslovakia proved mutually beneficial, but the Czechoslovak government's construction of a new nuclear power station after the early 1980s in Temelín, in southern Bohemia near the Austrian border, provoked continuing friction with Austria.

Another case of the failure of international cooperation on a matter of critical environmental and economic importance for the region was the Gabčikovo-Nagymaros Barrage System, a project initiated in 1977 by Czechoslovak and

Hungarian communist authorities to control the flow of the Danube between Bratislava and Budapest and to generate hydroelectric power. Faced by the serious environmental problems that were likely to ensue and by rising public opposition, the Hungarian parliament eventually voted to withdraw from the project in October 1989. The Czechoslovak authorities continued with their section of the project despite Hungarian opposition. In May 1992, Hungary terminated its treaty with Czechoslovakia for the project and dismantled the already built part of the system on the Hungarian side near Visegrad. In April 1993, Hungary and Slovakia submitted the disputed case to the International Court of Justice in The Hague. In the decision issued on 25 September 1997, the Court recognized the significance of the integrity of the ecosystem while finding that Hungary had violated the 1977 treaty by abandoning the project and Czechoslovakia had breached it by carrying out its "provisional solution."

After 1989, the new democratic governments of CEE pledged to improve environmental quality by implementing higher standards of environmental protection, more effective economic instruments, and new strategies of sustainable economic development. For Austria and the postcommunist governments alike, accession to the European Union (EU) has raised even further the environmental standards that must be met. Austria, which joined the EU in 1995, and most of the neighboring CEE states, which joined nine years later in May 2004, must deal now with a whole new body of environmental regulations coming out of Brussels.

Today's situation is completely different from that of just sixteen years ago, at the end of 1980s, when the region faced deep political, economic, and environmental crises. How did it happen that during such a short period the situation has undergone such dramatic change for these countries? What were the major forces behind those changes? Are the changes sustainable? How have these countries constructed the policies and institutions that facilitate such dramatic changes? What are they bringing to the EU? What can the CEE countries learn from the experience of closer and more recent EU members such as Austria? Those and many other questions set the stage for an international academic conference at the end of September 2002: "The Environment and Sustainable Development in the New Central Europe: Austria and Its Neighbors," organized at the Hubert H. Humphrey Institute of Public Affairs at the University of Minnesota by the Center for Austrian Studies in cooperation with the Center for Nations in Transition.

This book is a collection of about one third of the papers presented at the conference, carefully selected by the editorial team. The main criteria for this selection were the novelty of ideas and findings coming from the research papers and their often interesting methodologies, revealing unknown aspects of development in Central Europe.

This volume begins with an overview of the problems of sustainable development in Robert Wilkinson's provocatively titled essay "From Communism to Climate Change: The Sustainability Challenge and Lessons from Central Europe." Working for several years as an academic administrator in the heart of Central Europe, Budapest and Prague, Wilkinson was not only an attentive observer but also a critical analyst of development in this region. Contrary to the common belief in the West, he argues that Central Europeans, despite their tremendous environmental, economic, and social problems, "have much to teach the world." In particular he points to the lessons from Central Europe's transition in the changing role of women and governmental priorities related to the environment.

In Part One, the issue of **The Environment as Policy Priority** is explored by the American-Polish team of Sandra Archibald and Zbigniew Bochniarz in their chapter on "Assessing Sustainability of the Transition in Central European Countries: A Comparative Analysis." The authors challenge the popular opinion in environmental circles, promoted by Nobel Prize laureate Joseph Stiglitz, among others, that rapid economic liberalization of countries in transition undermines their economic and social sustainability. Archibald and Bochniarz conduct a careful trend analysis for twenty-five countries in transition and apply the econometric model of the Environmental Kuznets Curve to argue that fast liberalization did not impede the sustainability of the ten Central and East European countries (CEEC10) selected for accession to the EU. On the contrary, rapid liberalization was one of the major sources of their significant improvement in economic, social, and environmental performance. Early liberalizers from these ten countries confirmed the hypothesis that fast liberalization, accompanied by carefully designed institutional reform, human capacity building, and the application in environmental policy of certain market-based-mechanisms (MBM) that mobilize local resources—mostly pollution and user fees and product charges—could contribute significantly to sustainable development. Making the environment a top priority for transformation policies did not cause any harm to their economies. This is an important lesson, applauded in OECD and World Bank publications, that is still not understood by many political and business leaders.

Significant improvements in environmental, economic, and social performance paved the way for eight of the CEEC10 to join the European Union on 1 May 2004, and for the other two, Bulgaria and Romania, to be potentially ready for EU membership by 2007. The question, however, is whether the positive trends generated by radical transformation and industrial restructuring and driven mostly by national policies and local resources will be sustained after accession to the EU and their progress toward sustainability secured. This and several other questions are the subject of the next essay, contributed by American authors Stacy D. VanDeveer and JoAnn Carmin, on "Sustainability and EU Accession: Capacity Development and Environmental Reform in Central and Eastern Europe." The authors explore one of the least publicized aspects of EU enlargement, the environmental implications of "Europeanization" of accession countries' institutions, policies, expectations, and beliefs. They conclude that a successful enlargement process will depend on meeting three challenges: (1) developing appropriate institutional, human, organizational, and technological capacities for the environment, (2) securing sufficient financial resources, and (3) sustaining positive trends and further improving the quality of the environment. It is worth noting that these challenges face not only all new and future EU members, but existing members as well. Among some EU members, there is insufficient compliance with environmental standards and laws. Overall, the quality of the environment in the EU is not improving or is improving only very slowly, and it is at great expense that members meet all environmental regulations, particularly those based on the "best available technology" (BAT). EU members face the additional challenge to learn from CEEC10 experiences, particularly with implementation of more MBM (instead of BAT) that can significantly reduce the cost of policy implementation for industries and boost innovative institutional reforms.

The sustainability issues facing both CEEC10 and established EU members are further explored by Edward M. Bergman within the framework of distinct localities and regions in his research, focusing on "Sustainability of Clusters and Regions at Austria's Accession Edge." The author starts by challenging traditional definitions of "sustainability" and "clusters" in the search for answers to such basic questions as what "sustainable cluster" means, and when does a cluster reach sustainability. For Bergman, cluster sustainability means the continued generation of innovation and a secure competitive position within a geographically bounded business network. These are the critical issues for local and regional prosperity in growing competition in more and more globalized economies. This is why, he argues, in contemporary development theory and the policy of EU members, the concepts of industrial clusters and decentralized regional development capacities play such an important role, contrary to the previous theories and practices of CEE countries under communist rule, which focused on centralized national economies. This is an important lesson for the accession countries which can learn from the positive Austrian and EU experiences how to boost and sustain economic development by promoting industrial clusters, particularly for small and medium enterprises, and building strong regional economies.

The second part of the book focuses on **The Economics of Sustainable Development** and contains two interesting papers on innovative institutional and policy reforms in the Czech Republic and Poland. The Czech team, composed of Jiřina Jílková and Thomas Chmelik from the University of Economics in Prague, presents an analytical picture of the development of "Greenhouse Gases Emissions Trading in the Czech Republic." This has been a hot issue since the Kyoto Protocol on the reduction of greenhouse gases (GHG) triggered global debate and several surprisingly radical twists in policy positions, particularly in the United States and Russia. The authors present a well-documented background against which to analyze energy intensity and associated pollution trends in the Czech economy and other transitional economies and to evaluate the development of policies on global climate change. They explain major policy drivers in addition to the EU accession requirements. Having laid these foundations, they present a variety of policy options to tackle GHG reduction. The options include flexible mechanisms preferred by economists and business leaders, particularly emission trading and joint implementation, accompanied by air pollution charges, tax reforms with subsidies for renewable energy sources and conservation, and regulatory instruments.

The authors concentrate their attention on two basic mechanisms as the most critical for successful GHG reduction: emissions trading, as a main pillar; and joint implementation (JI), as a pilot project for climate change policy and a complementary mechanism. Due to the fact that the Czech Republic is a leader in JI among the CEEC10 with significant potential for further GHG reduction, the analytical contribution of Jilková and Chmelík offers an attractive lesson for other countries and suggests good practices to follow. What is probably even more significant is the clear message from the Czech authors that climate change is treated seriously in the Czech Republic as an important policy and business issue, which facilitates comprehensive responses from both government and business interests and has already attracted domestic and foreign investors.

The Polish team of Olga Kiuila and Jerzy Śleszyński of Warsaw University also has an important message to share with their research on "Ecological Reform in the Tax System in Poland." The authors, following some of the promising applications from Scandinavian countries, propose a comprehensive tax reform that will not increase the tax burden but will contribute to significant improvements in environmental, economic, and social sustainability. This is a great challenge, given the growing aversion to new taxes all over the world. Despite the fact that Poland developed the most comprehensive and effective emissions fee system in the 1990s, the external motivation from EU countries to develop further the environmental tax system encouraged research and preparation of a new tax reform. Internal factors, too, were crucial. These included significant institutional weakening of the existing emissions fee system after the implementation of the 1999 administrative reform and the need to raise about USD 30–50 billion for implementing EU environmental requirements.

The main goal of the proposed tax reform is to reach a double dividend: improve both environmental quality and economic efficiency. The authors apply a complex general equilibrium model to simulate six different scenarios of tax reform. The research reveals that three of the proposed scenarios could secure long-term, environmentally sound economic growth with significant reduction of unemployment. Now the challenge is to convince the policy makers in Poland to implement it, as their counterparts have already done in some other EU countries.

The third part of the book is devoted to **Water Policies and Institutions** and their sustainability. It begins with two contributions devoted to environmental disasters, which also treat broader, related concerns: institutional sustainability, popular values and attitudes, traditions and readiness to change, and the role of environmental disasters in shifting paradigms. The first contribution comes from the Czech scholar and environmental activist Václav Mezřický and the second from an international team of American and Hungarian scholars including Jim Perry, Eszter Gulácsy, and László Pintér.

In the first essay, entitled "The Czech Republic: From Environmental Crisis to Sustainability," the author uses the case of the 2002 flooding, the worst in the last 150 years, as a starting point for his analysis of the concepts of sustainability over several decades. Although there was a large body of research reports, policy documents, and publications raising environmental awareness and educating policy makers in the Czech Republic, he argues that neither the concepts nor the policies implemented took into account the possibility of such a major environmental disaster. The tragic events of 2002 shocked the country and have heightened environmental awareness significantly, opening a window of opportunity to correct previous concepts and redesign institutions and policies that will be based on correctly defined sustainability principles and will apply proper methodological concepts, e.g., environmental space or footprint. Although policies and institutions are very important for sustainability, Mezřický concludes that without fundamental changes in the values and attitudes of individuals, significant breakthroughs cannot happen.

The second contribution in this section focuses on "The Tisza/Tisa Transboundary Environmental Disaster: An Opportunity for Institutional Learning." Here again, as in the case of the 2002 Czech flooding, the authors have researched an environmental disaster, the cyanide spill from a gold mine near Baia Mare in Romania, which poisoned the Tisza/Tisa River and caused catastrophic damage downstream in Hungary. In this chapter they analyze the causes, identify institutional and policy deficiencies, and propose policy modifications. They find that the Tisza/Tisa River is a typical case of a transboundary waterway in the transitional countries and for that reason should be treated as an important lesson for all of them. From the standpoint of environmental management, they argue that technical knowledge alone is insufficient for sound decision making, which requires also basic knowledge of the institutional, local, and national cultures affecting the individual decision makers. These individuals, shaped often by decades of anonymous, hierarchical, centralized, and nontransparent decision making in the old system, were then expected to be accountable, participatory, and transparent. This presents a serious challenge to the individuals and the society. Besides the inherited deficiencies at the individual level, the disaster revealed many gaps in institutional arrangements and policy measures at the national, regional, and local levels in Romania and Hungary that could have parallels in other countries.

One of the solutions for filling the existing gaps in institutional and policy arrangements in CEE is to follow EU legislation, as Austria has done since 1995. According to the essay by Wilhelm R. Vogel from the Austrian Federal Environment Agency, "Austria and the EU Water Framework Directive," the major challenge was faced not at the beginning of membership, but rather with the 2000 Water Directive requiring restructuring of water management in all member states and the introduction of new quality standards and new approaches to management. Vogel argues that the lessons learned by Austria before accession and during the implementation of new EU regulations could be beneficial to the CEEC10 in the near future.

Managing environmentally sound transboundary resources is discussed again in the essay by James B. Dalton on "The Western Bug River: UNECE Pilot Project." The significance of this shared resource, the Western Bug River, derives from the fact that this river marks the NATO border, and since May 2004 also the EU border. Pollution from upstream is discharged directly into the Western Bug and its tributaries in Ukraine and in Belarus, which can endanger the environmental security of Poland, a NATO and EU member. In his essay Dalton explores both scientific data and current institutional arrangements on both sides of the river for preventing such pollution, based on his own research and the United Nations Economic Commission for Europe (UNECE) Interim Report on the Water Convention Pilot Project Programme on Transboundary Rivers. Dalton concludes that the agreements signed by Poland and Ukraine are a good foundation for preventing environmental disasters and bringing stability to the river's management, so long as they are fully implemented.

The final essay on water resources, "Wastewater Treatment in the Postcommunist Danube River Basin," contributed by the Slovak scholar Igor Bodík, also focuses on a transboundary river. This essay deals with the whole Danube basin, beginning with a presentation of different aspects of the economic, social, and geographical characteristics of all thirteen Danube River Basin (DRB) countries. Then Bodík moves to a brief description of water supply and use patterns, which builds a foundation for wastewater analysis. This whole analytical work leads Bodík to address policy and institutional issues and recommendations.

The fourth and last part of the book is devoted to the troublesome issue of **Agriculture and Rural Development** in CEE. It begins with Anthony J. Amato's chapter on the eastern Carpathian region of Ukraine, entitled "'Thinking Unlike a Mountain': Environment, Agriculture, and Sustainability in the Carpathians." This interdisciplinary study starts with a historical analysis of the Galician Hutsul region and then moves through geographical and ecological characteristics to present historical trends in agricultural activities. It tells the story of almost two hundred years of human pressure on a unique natural environment, leading to its decline and the disappearance of species characteristic of the region, such as European bison and bears. The Soviet legacy for regional agriculture and forest management is shrinking biodiversity, which was exacerbated after Ukraine's independence in 1991 by the introduction of a market economy without appropriate institutions and policies. Despite all these setbacks, Hutsul agriculture, Amato argues, exhibits strongly resilient characteristics, durability, persistence, and high energy efficiency, which may present a chance for survival in the twenty-first century to this region and its unique environment and culture.

Survival and sustainability issues are also discussed by Slovak scholar and community development leader Slavomíra Mačáková. Her essay, "New Approaches to Sustainable Community Development in Rural Slovakia," focuses on the country's poorest, most marginalized and vulnerable ethnic group, the Roma population. The demographic situation shows that Roma make up 8 to 9 percent of the Slovak population, one of the highest shares in Europe. This minority suffers from a very low level of education, poor health conditions, and high long-term unemployment. Yet, despite all the problems and obstacles that lead many Roma to hopelessness and despair, there is a chance, argues Mačáková, based on her three years of successful experience in the Spiš region, to empower the Roma population and move them step by step out of poverty toward a more sustainable and prosperous life. This is a very special and most valuable object lesson, not only for the other nations in transition but also for developed nations facing problems with their minorities.

Another aspect of rural sustainability is explored by the Czech-Canadian team of Antonín Vaishar and Bryn Greer-Wootten in their chapter devoted to "Sustainable Development in Moravia: An Interpretation of the Role of the Small-Town Sector in Transitional Socioeconomic Evolution." The authors begin by analyzing the theoretical underpinnings of the "new regionalism" that is emerging along with a decline in the role of nation states due to globalization and the growing significance of sustainability concerns. Then they offer a comprehensive analysis of the small-town sector (settlements with fewer than 15,000 inhabitants) in Moravia, stating that there is a future for these towns in facilitating sustainable rural development. One of the interesting conclusions they advance is that small towns have a "mediating influence" between urban and rural areas that will affect the future Czech population structure and its sustainability.

Finally, the sustainability of rural areas in Hungary is assessed by the environmental leader and educator Judit Vásárhelyi in her essay called "Building Local Sustainability in Hungary: Cross-Generational Education and Community Participation in the Dörögd Basin." The author first describes the country's population distribution, which is characterized by a significant concentration of urban communities around the capital city, and outlines its consequences. Then she examines the environmental problems threatening local communities. Based on her experience and research, Vásárhelyi argues that through participatory processes a local community can learn about environmental threats and generate sound responses. The case of the Dörögd basin illustrates her arguments.

This fourth part of the book, devoted to rural communities, shows that despite the well-recognized crises of rural areas, there are still many promising cases that illustrate opportunities to overcome crisis situations and exemplify good practices. The essays in this section, as well as in the previous sections of the book, offer weighty yet optimistic findings. Crises and even serious environmental disasters need not discourage people from setting ambitious goals to improve their quality of life and to mobilize their own resources. The citizens of CEE show that human creativity in the quest for innovative solutions is a powerful force that can overcome economic and technological barriers and—even more challenging— barriers of inertia and conservative thinking. The hard-learned lessons in CEE reaffirm that there is much that both developed and developing countries can learn from this region. If this book helps to propagate such lessons, it will give great satisfaction to its authors and editors.

Chapter 1

FROM COMMUNISM TO CLIMATE CHANGE
The Sustainability Challenge and Lessons from Central Europe

Robert Wilkinson

Central Europeans have much to teach the world. Nations, communities, corporations, and citizens are confronting profound changes in global environmental, political, economic, and social systems. As global systems change, we respond in different ways—some effective and positive, some not. This short article posits the notion that Central Europe is a place to look for an understanding of how to process profound and fundamental change in effective, civilized, thoughtful ways. The world should, I argue, study examples provided by the people of this region for clues that may help bring about a transition to sustainability.

There is no obvious connection between the ideology behind a political/economic system and a physical process occurring in the atmosphere. I suggest that the real challenges posed both by ideology and global environmental challenges are *people problems.* That is, they are problems largely of our own making and they are problems that can best be resolved by thoughtful people committed to civilized change. The science of climate change is reviewed briefly below. But I will argue that science is not the hard part. Sustainability will require changes in our thinking: our notions of right and wrong, good and bad. It will require a new level of dialogue and decision making in the long-term public interest.

In the past century, Europe experienced serious consequences from an exaggerated enthusiasm for ideologies: nationalism, fascism, communism, and now perhaps capitalism. Rather than dwell on communism, let me instead give you a glimpse of what I saw in Central Europe and throughout the Soviet Union as the Soviet system collapsed a little over a decade ago. It is the *transition* from that system to an entirely new one that I wish to focus on. Specifically, it is the way people were able to process this change in what can only be described as a remarkable—indeed elegant—process.

My connection to Central Europe stems from a special opportunity I had in 1990 to set up a new graduate program in Environmental Science and Policy at the newly formed Central European University. I had the pleasure of recruiting brilliant young scholars in every country of Central Europe—from Albania and the countries of what was then Yugoslavia (as it was coming apart) in the south, up through the "center" of Central Europe to the Baltic states in the north. All of the countries of the former Soviet Union were included, so I also recruited students across Belarus, Ukraine, and Russia to eastern Siberia, and down through Central Asia as well. It was a unique time in history. When I started, the Soviet Union was not "former" and Czechoslovakia was still one country. The wall was just down and Germany had been re-united. Yugoslavia was in the process of coming apart.

My job was to figure out the key environmental issues throughout the region, set up a brand new graduate program and curriculum, and then recruit scholars from every country. As I traveled throughout the region I found extraordinary beauty in old European cities, forests, mountains, agricultural lands, and communities. I also witnessed the problems first-hand. The challenges were immense. At Chernobyl, I was told I had five minutes to observe the crumbling sarcophagus, whereupon I would be at the maximum "safe" dose of radiation. The people of Ukraine and Belarus are still coping daily with the effects of the "event" at the nuclear facility. At the Aral Sea I witnessed the destruction of a once-rich fishery and ecosystem. Again, the people are living with this disaster today. To the west, in Central Europe, I visited "secret" towns that were not on the maps. They were the source of "yellow cake" for bombs. Open ponds held radioactive waste. The list of environmental problems was long. Coal mines, power plants, and steel mills polluted the air and water throughout the region. This is the image that many in the West saw on the covers of magazines like *The Economist* and *National Geographic*. What many failed to see was the strength of the people and their desire to fix the problems and rebuild their communities—physical, human, and ecological.

Throughout the region, in every country, I was inspired by people's ability to think clearly and creatively about the changes that were needed and ways to get started. I interviewed political leaders, dissidents, scientists, students, and many others to learn the nature and extent of the specific environmental issues facing the region. It was a young woman in Prague who provided the most interesting and insightful response. Foregoing the usual answers to my questions regarding problems—air pollution, forest death, water pollution and so on—she looked at me with serious and thoughtful eyes and said: "It is us, of course. It is our way of thinking. That is our environmental problem. That is what we must change."

People like this young woman in Prague played a major role in the profound transition of Central Europe in the late 1980s and early 1990s. I have no scholarly research upon which to base this assertion, but I make it with confidence nonetheless: Women played a critically important and unique role in the changes that occurred in Central Europe and the former Soviet Union. (This does not mean, of course, that women played an exclusive role. I do believe, however, that their contribution to this important chapter of history is underappreciated.) I had the privilege to meet many of these brave women in every country of Central Europe, and throughout the former Soviet Union. Their strength and courage was, I believe, a key part of the reason that the transition occurred as it did. They often faced down authorities with little more than a passionate concern for their children's health and a clear sense of the need for change. Their actions were not without substantial risk. Yet, they prevailed. The authorities knew they were right. And they also realized that these women were serious about the changes that were needed. In community after community, country after country, I met these women and listened to their stories. Their impact is a piece of scholarship that should be pursued by someone far more qualified than me. The history of the transition of the Soviet system, and in particular the role of women and environmental issues, is an important story. It is one of the important lessons from Central Europe, and it is one that should not be lost.

As we contemplate major challenges confronting the world a decade after the transition in Central Europe, it is clear that in the way Central Europeans handled profound changes there are lessons and valuable insights for a world facing the challenge of sustainability. Of these challenges, climate change due to global warming is perhaps the most serious. The scientific debate over the existence and seriousness of the issue is fairly well understood, except in some scientifically challenged quarters. A quick review of the science is merited, but it is presented only to consider the opportunity we have to apply lessons about change processes.

There is broad scientific agreement that global warming is occurring and that climate change and variability pose important challenges. Scenarios developed in 2001 for the third assessment by the Intergovernmental Panel on Climate Change, the international effort examining the science of climate change, indicate that in the period from 1990 to 2100, surface temperatures (averaged globally) will increase by 1.4 to 5.8 °C (2.5 to 10.4 °F) relative to 1990. Global mean sea level is projected to rise by 0.09 to 0.88 meters (3.5 to 35 inches), primarily due to thermal expansion and glacial and ice sheet melting.[1]

The U.S. National Research Council's (NRC) Committee on the Science of Climate Change confirmed in its report to the current U.S. administration *(Climate Change Science: An Analysis of Some Key Questions)* that "Greenhouse gases are accumulating in Earth's atmosphere as a result of human activities, causing surface air temperatures and subsurface ocean temperatures to rise. Temperatures are, in fact, rising."[2] The NRC committee's report confirmed the work of the Intergovernmental Panel on Climate Change (IPCC). The committee found that the IPCC report "is an admirable summary of research activities in climate science."[3]

The U.S. government also has clearly acknowledged the role people play in causing this change:

Humans are exerting a major and growing influence on some of the key factors that govern climate by changing the composition of the atmosphere and by modifying the land surface. The human impact on these factors is clear. The concentration of carbon dioxide (CO_2) has risen about 30% since the late 1800s. The concentration of CO_2 is now higher than it has been in at least the last 400,000 years. This increase has resulted from the burning of coal, oil, and natural gas, and the destruction of forests around the world to provide space for agriculture and other human activities. Rising concentrations of CO_2 and other greenhouse gases are intensifying Earth's natural greenhouse effect. Global projections of population growth and assumptions about energy use indicate that the CO_2 concentration will continue to rise, likely reaching between two and three times its late-19th-century level by 2100. This dramatic doubling or tripling will occur in the space of about 200 years, a brief moment in geological history.[4]

At the time of the *Environment and Sustainable Development in the New Central Europe* conference, I had just returned from meetings in Hungary addressing the potential impacts of climate change on Lake Balaton. In the preceding weeks, Central Europe experienced flooding in Austria, the Czech Republic, Germany, and elsewhere that was described as a "hundred-year" event. Similar problems occurred to the east in southern Russia and other parts of the region. Lives were lost. Damages were in the billions of U.S. dollars. When I was there in the fall, there was considerable discussion of climate change and the recognition that increases in precipitation and flooding are potential impacts associated with global warming. In short, Central Europeans were taking the threat seriously. Whether these floods were in fact a direct result of a changing climate is not the point. More crucial is that if climate change brings more events like the floods of 2002, it will be a serious problem. The Central Europeans I met with were asking a sensible question: What do we need to change to address *this* problem?

Like the Soviet system in decades past, it may appear to many of us at present that the worldwide economic system and energy infrastructure that are causing greenhouse gas emissions are unchangeable. Certainly, we might argue, they cannot be changed in a short period of time. Not many people would have predicted, even as late as the mid 1980s, that the Soviet system would be completely transformed—and eliminated—at the end of that decade. Yet, a huge political-economic system changed fundamentally in a matter of several years. It is astonishing to consider how thoughtfully all parties dealt with such profound changes. The scale and depth of the changes experienced by Central Europe were far greater than those projected in any scenario contemplated to deal with climate change. The systems through which we realize our productivity, mobility, and basic needs must change to avert serious climate change. As Central Europe has demonstrated, such change is both possible and desirable.

Our challenges are largely of our own making, from ideologies to environmental problems. We need to learn how to transition our systems—economic, political, and social—to sustainable models. Central Europe has much to teach the world about transitions toward sustainability. New policy approaches are emerging in response to a recognition of the need to restore and protect the environment while achieving productivity and profitability.

Central Europe taught the world by elegant example how economic and political systems can change for the better. As the young woman patiently explained to me: "The problem is us, of course. It is our way of thinking." Sustainability will require us to learn from the talented and thoughtful people of Central Europe. Let us hope we recognize this soon. As the Hopi elders remind us, "We are the people we have been waiting for."

Notes

1. Intergovernmental Panel on Climate Change (IPCC), *Climate Change 2001: The Scientific Basis. Contribution of Working Group I to the Third Assessment Report of the Intergovernmental Panel on Climate Change*, ed. J. T. Houghton, Y. Ding, D. J. Griggs, M. Noguer, P. J. van der Linden, X. Dai, K. Maskell, and C. A. Johnson (Cambridge: Cambridge University Press, 2001).
2. National Research Council, Committee on the Science of Climate Change, *Climate Change Science: An Analysis of Some Key Questions* (Committee on the Science of Climate Change, Division on Earth and Life Studies) (Washington D.C.: National Academy Press, 2001). National Academy Press, http://www.nap.edu
3. Intergovernmental Panel on Climate Change, http://www.met-office.gov.uk/sec5/CR_div/ipcc/wg1/
4. National Assessment Synthesis Team, *Climate Change Impacts on the United States: Report for the United States Global Change Research Program* (Cambridge: Cambridge University Press, 2001), 12. http://prod.gcrio.org/nationalassessment/

Part One

THE ENVIRONMENT AS POLICY PRIORITY

Chapter 2

ASSESSING SUSTAINABILITY OF THE TRANSITION IN CENTRAL EUROPEAN COUNTRIES

A Comparative Analysis

Sandra O. Archibald and Zbigniew Bochniarz*

Introduction

Since the late 1980s, the Hubert H. Humphrey Institute of Public Affairs at the University of Minnesota has been involved in policy-oriented research, institutional design for sustainable development, and the reform of management and economic education in seven Central and East European (CEE) countries. After conducting the bulk of the work in the 1990s, we became interested in examining whether the radical reforms introduced in those countries contributed to breaking negative economic, social and environmental trends in the CEE countries. Did economic reform help or hinder those countries in developing economic, social, and environmental sustainability?

The long-term goal of this project is to examine whether trends in environmental pollutants and social indicators demonstrate improvement in environmental quality and quality of life, i.e., more sustainable development in the CEE countries since these efforts were initiated. Economic literature on economic development posits an inverted U relationship between environmental quality and

---

* From the Center for Nations in Transition, Hubert H. Humphrey Institute of Public Affairs, University of Minnesota. The authors would like to acknowledge significant contributions by the Humphrey Institute graduate research assistants Luana E. Banu, Andrea Cutting, and Michael H. Hamann.

per capita income in countries as they develop. According to the environmental Kuznets curve (EKC), environmental pollutants or pressures increase with growth. After some critical turning point when a certain level of income is attained, there is a decline in environmental pressure or pollutant emissions.[1]

Research we conducted previously using data from the period from 1990 to 1996 provided some empirical evidence in support of the environmental Kuznets curve hypothesis for six CEE countries (Poland, Czech Republic, Slovakia, Hungary, Romania, and Bulgaria). Interestingly, the turning point was attained much earlier than had been expected. Depending on the model used and pollutant studied, the turning point occurred between $2,800 and $4,500 gross national product (GNP) per capita, whereas conventional knowledge indicated that pollution would not decline until after a per capita income of $5,000–8,000 was attained. Our analysis indicated the critical importance of examining the progress in economic transition by including measurements of structural changes in the economy and rapid liberalization as measured by openness in trade and investment.

Assessing outcomes of institutional reforms and environmental and other policies designed to promote environmentally sound restructuring proved equally important for understanding the decline in pollutant emissions in the CEE countries. Our analysis showed that these reforms had a significant positive effect on the environment and could explain a significant share of the observed declines in emissions over the period. The analysis clearly pointed to the need for continued progress toward privatization, openness in trade, and improvements in environmental regulatory infrastructure and investments. We concluded that future increases in income, in the absence of further policy initiatives to assure environmentally sound development, could be expected to increase pressure on the environment in the near term and the medium term. This provided preliminary evidence that a rise in demand for environmental quality was likely to occur for most pollutants at lower income levels than previously believed. It also confirmed our preliminary hypothesis that progress in the transition to a market economy and civic society in CEE countries contributes to sustainable development in the region.

Encouraged by our preliminary results, we continued our investigation of sustainability of economies in transition by introducing additional institution reform measures and expanding the analysis to twenty-five countries of CEE and the New Independent States (NIS).[2] We were also motivated by a polemic between Joseph Stiglitz[3] and a group of Polish economists, including Marek Dabrowski, Stanislaw Gomulka and Jacek Rostowski,[4] regarding the sustainability of transition.

Based on the analysis of transitional reforms and economic policies, particularly in China and Russia, Stiglitz claims that: "The rapid transition attempted in the FSU and Eastern Europe has led to increased inequality, increased poverty and reduced life expectancy.... Rapid liberalization did not lead to rapid growth."[5]

The Polish economists argue that by taking into account only two models of transition, Chinese and Russian, Stiglitz has overgeneralized the transition experience. In addition, they challenge several assumptions made by Stiglitz in the

Russian model. They conclude that the CEE countries' transition experience showed features of economic and environmental sustainability that could not be shown in the dichotomous China-Russia comparison. Masahiko Gemma's research also identified different patterns of transition in CEE countries, including the Baltic countries, versus the rest of the NIS.[6]

Because of the brevity of this paper, we are unable to present all of our results. Accordingly, we have mostly selected figures comparing the CEE4 and non-CEE4 to illustrate our findings. For this paper, the terms CEE4 and Visegrad (VG) are equivalent. Both refer to the Czech Republic, Hungary, Poland, and the Slovak Republic. As becomes obvious in the discussion of transition indicators, by all measures the CEE4 represent faster transitions to market economies and civil societies. Non-CEE4 can be understood to represent slower transitions.

Does the Speed of Transition Matter for Sustainability?

One of the primary issues in this polemic was whether the speed of transition to a free-market and open civil society matters for sustainability. We hypothesized that rapid liberalization might not necessarily exclude the possibility of achieving sustainable development. With over fifteen years of data from the region, we wanted to test this hypothesis empirically. We further suspected that public policy could affect the outcomes of economic transition. We hypothesized that countries experiencing faster economic transitions also passed the EKC turning point sooner and have achieved declining pollutant emissions and greater social and economic stability, thus contributing earlier to the economic, social, and ecological sustainability of the region.

Measuring the Pace of Transition: Four Indicators

In order to measure the speed of the transition process, we chose to characterize countries according to their degree of market liberalization, price stabilization, privatization, and financial institution reforms.[7] These four indicators determine the degree to which the markets of these countries were open to global trade and basic market institutions, such as private enterprise.

First Transition Indicator: Market Liberalization

We assessed progress in market liberalization in each of the selected twenty-five countries and ranked them according to the methodology developed by the European Bank for Restructuring and Development (EBRD). The following criteria were considered: complete price liberalization, full current account convertibility, and almost complete small-scale privatization.[8] Based on these three criteria, the twenty-five countries were divided into three groups:

- *Early Liberalizers:* Czech Republic, Estonia, Hungary, Poland, Slovak Republic, and Slovenia.
- *Late Liberalizers:* Albania, Armenia, Bulgaria, Croatia, FYR Macedonia, Georgia, Kazakhstan, Kyrgyzstan, Latvia, Lithuania, Moldova, Romania, and Russia.
- *Non-Liberalizers:* Azerbaijan, Belarus, Tajikistan, Turkmenistan, Ukraine, and Uzbekistan.

Second Transition Indicator: Economic Stabilization

The main criterion for assessing the progress of stabilization was reduction in the rate of inflation. A country is considered an early stabilizer if inflation fell to a rate below 50 percent by December 1994 and was sustained below that level in the following years.[9] This is a more stringent criterion compared with that of the EBRD, which does not require stabilization over time.[10] Based on this criterion, the countries fall into two groups:

- *Early stabilizers:* Albania, Croatia, Czech Republic, Estonia, Hungary, Latvia, Lithuania, Poland, Slovak Republic, and Slovenia.
- *Late stabilizers:* Armenia, Azerbaijan, Belarus, Bulgaria, FYR Macedonia, Georgia, Kazakhstan, Kyrgyzstan, Moldova, Romania, Russia, Tajikistan, Turkmenistan, Ukraine, and Uzbekistan.

Third Transition Indicator: Privatization

In assessing the progress in privatization, we applied EBRD criteria for early and late privatization to two different size categories of firms within each country. Category 1 describes countries based on the degree to which medium-sized and large enterprises have been privatized. In Category 1, the *early privatizers* group contains those countries that had privatized at least half (50 percent) of their medium-sized and large enterprises by the end of 1998. Category 2 depicts the degree to which privatization has taken place among a country's small-scale enterprises. In Category 2, the *early privatizers* are those countries that had privatized a minimum of 70 percent of their small-scale enterprises by the end of 1998. The results of applying these criteria follow.

Category 1: Medium and Large Enterprise Privatization

- *Early privatizers:* Armenia, Bulgaria, Croatia, Czech Republic, Estonia, FYR Macedonia, Georgia, Hungary, Kazakhstan, Kyrgyzstan, Latvia, Lithuania, Moldova, Poland, Russia, Slovak Republic, Slovenia.
- *Late privatizers:* Albania, Azerbaijan, Belarus, Romania, Tajikistan, Turkmenistan, Ukraine, and Uzbekistan.

Category 2: Small Enterprise Privatization

- *Early privatizers:* Albania, Croatia, Czech Republic, Estonia, FYR Macedonia, Hungary, Kyrgyzstan, Latvia, Lithuania, Poland, Romania, Russia, Slovak Republic, Slovenia.
- *Late privatizers:* Armenia, Azerbaijan, Belarus, Bulgaria, Georgia, Kazakhstan, Moldova, Tajikistan, Turkmenistan, Ukraine, and Uzbekistan.

Fourth Transition Indicator: Reform of Financial Institutions

The evaluation of the progress in reform of financial institutions in these twenty-five countries is based on the rating system developed by the EBRD, yielding the following results:

- *Advanced reformers* (score of 3–4): Croatia, the Czech Republic, Estonia, Hungary, Lithuania, Poland, and Slovenia.
- *Progressing reformers* (score of 2–2.99): the rest of the CEE countries and all but three members of the NIS.
- *Lagging behind with reforms* (score of 1–1.99): Belarus, Russia, and Uzbekistan.[11]

Evaluating Trends: The Impact of Transition on Sustainability

Development affects all areas of society, and truly sustainable development must be sustainable in each of those areas. To look broadly at outcomes of transition in the CEE countries, we examine the impact of transition in the economic, social, and environmental arenas. There are, in theory, many ways of measuring the degree to which change is sustainable. Because we are limited to the data we have available for the countries under consideration, our choice of measures is somewhat restricted.

Economic Sustainability

There are many criteria for evaluating economic sustainability. We examine the effects of transition on per capita income, on relative size of the service sector, and on participation in global trade.

Per Capita Income. Two useful measures of economic sustainability are nondeclining income per capita and nondeclining genuine (net) savings, also called the "green" Net National Product.[12] Using Gross National Product (GNP) alone measures income apart from environmental degradation. "Green" Net National Product, on the other hand, adjusts for the failure of GNP alone to capture environmental degradation as a cost to the economy. Unfortunately, data are not yet sufficient to calculate this adjustment. For that reason, trends in economic sustainability are measured in this paper using GNP per capita in constant 1995 U.S. Dollars (USD).[13]

We examined the degree to which stable or rising per capita income had been achieved in light of the four transition indicators (liberalization, stabilization, privatization, and financial sector reform). We found that all contributed significantly to gains in per capita incomes and thus to sustainability. Despite a one-year decrease and roughly three years of stagnation at around USD 3,000 per capita GNP, early liberalizers, early stabilizers and advanced reformers have experienced sustained annual increases of GNP per capita since 1995. These countries reached over USD 4,000 per capita GNP in 1999.

There is a significant income gap between the CEE4, or VG countries and the non-CEE4 group, as illustrated by Figure 2.1. Unlike the CEE4, the GNP for the rest of the countries in transition decreased from about USD 2,500 (1990) to about USD 1,600 (1996) and remained stagnant at that level until the end of the period of analysis. Similar results were achieved by late stabilizers.

Late liberalizers experienced an absolute loss in per capita income over the same time period, from about USD 3,000 to about USD 1,800, a fall of 40 percent. Losses were even larger, in percentage terms, for non-liberalizers, from USD 1,800 to USD 900, a decline of 50 percent (see Figure 2.2). Similar conclusions are shown in Figure 2.3, which compares progressing reformers with those lagging behind. In general, countries characterized as late liberalizers, non-liberalizers, late stabilizers, progressing reformers, or lagging behind did not make headway toward reaching economic sustainability, as measured by per capita income, until 1999. As a consequence they did not experience any significant gains from the transition process other than halting economic decline, in most cases.

More problematic is how to interpret the impact of the privatization process on per capita income. In general, privatization did not produce obvious gains in income. However, within each category of enterprise size, early privatizers performed better than did late privatizers. By the end of 1999, early privatizers of

Figure 2.1 GNP per Capita

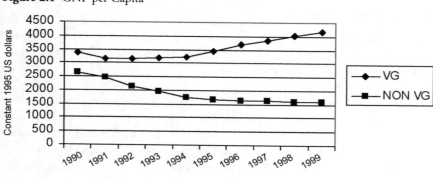

Figure 2.2 GNP per Capita—Liberalizers

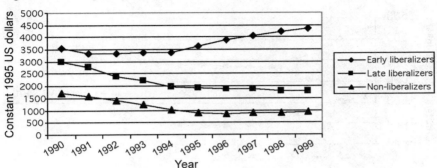

Figure 2.3 GNP per Capita—Financial Reformers

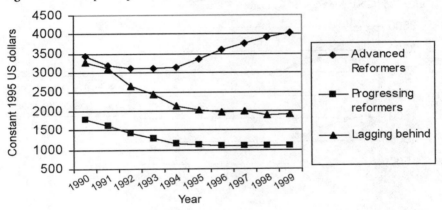

both large and small enterprises demonstrated almost three times higher average GNP per capita than the corresponding late privatizers in each size category (Figures 2.4 and 2.5). Although Category 2 (small enterprise) early privatizers performed a little better than Category 1 (medium and large enterprise) early privatizers, by 1999, neither group had again achieved the income level they had enjoyed before embarking on transition (USD 3,200–3,300 per capita). For both categories of early privatizers, the increases in GNP per capita after 1995 were negligible.

One major reason for the better performance of privatization of smaller firms (Category 2), as opposed to that of medium and large businesses, was that these enterprises were forced by tight budget constraints to restructure and to attain efficiency from the start. In the case of medium and large enterprises, privatization was introduced mostly through a variety of mass privatization schemes, including vouchers, which led neither to rapid restructuring of management nor to raising new capital. As a result, larger enterprise privatization (Category 1) did not accomplish the rapid efficiency gains achieved by smaller enterprise privati-

Figure 2.4 GNP per Capita—Category 1 Privatizers

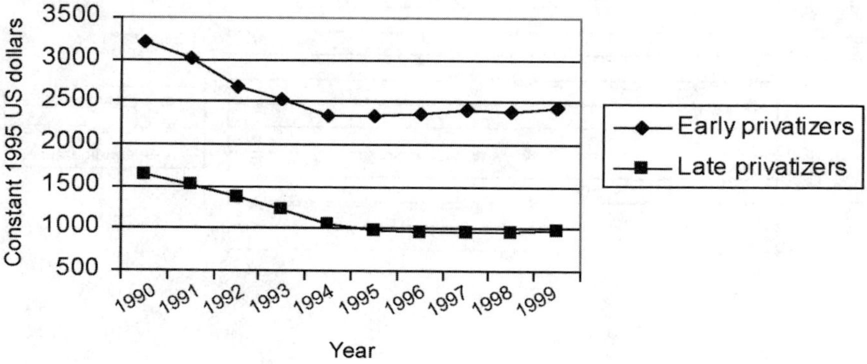

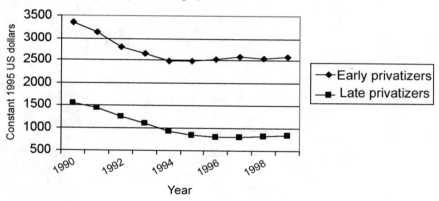

Figure 2.5 GNP per Capita—Category 2 Privatizers

zation (Category 2). In the case of late privatizers, by 1996 they enjoyed a GNP of only USD 800–1000 per capita, down from about USD 1,500–1,700 per capita in 1990. This same group remained stagnant after the mid 1990s, neither rising nor falling further.

Structural Change and a Growing Service Sector. Economic sustainability depends on a country's ability to bring about the structural changes necessary to be competitive on the world market yet meet the social welfare needs of its society. One measure of wealth and competitiveness in the world market is the share of the service sector as a percentage of GDP. As countries adopt more efficient technologies and liberalize market systems, an increasingly smaller share of GDP will be made up of basic industry. The postcommunist countries entered the transition period with a relatively low contribution (30–40 percent) by the services sector to the GDP due to long-term preferential policies toward the heavy industries. This low percentage adversely affected their ability to meet the challenges of modern society and global economy. Thus, the share of the service sector or, conversely, the share of manufacturing as a percentage of GDP serves as our principal structural change indicator.

As shown in Figures 2.6 and 2.7, by 1996, the service sector surpassed 60 percent of the GDP among early liberalizers and the CEE4. After 1996, those countries sustained modest growth in the service share of GDP. The early stabilizers (Figure 2.8) and advanced reformers (Figure 2.9) also were able to reach and sustain a 60 percent share of the service sector in GDP after 1997–98. Despite significantly faster growth of services among early privatizers than among late privatizers in each size category, the early privatizers reached only a 58 percent share in GDP in 1997 (Figure 2.10) and sustained minimal growth from then on. In general, the data indicate that the leaders in market liberalization, stabilization, financial reform, and privatization were able to perform structural reforms faster and have managed to sustain them better over the long term.

Participation in Global Trade. The economic sustainability of a country depends on its ability to compete and trade globally on the world market. One typical feature of centrally planned economies was low participation in world trade.

Figure 2.6 Contribution of Services to GDP—Liberalizers

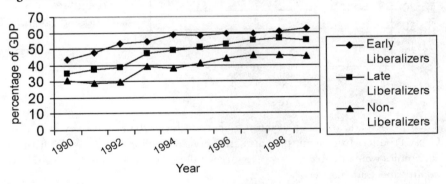

Figure 2.7 Contribution of Services to GDP

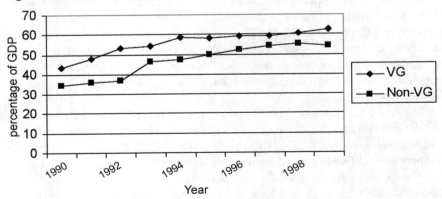

Figure 2.8 Contribution of Services as Percent of GDP—Stabilizers

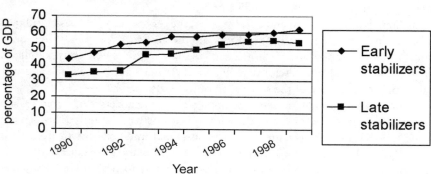

Figure 2.9 Contribution of Services to GDP—Reformers

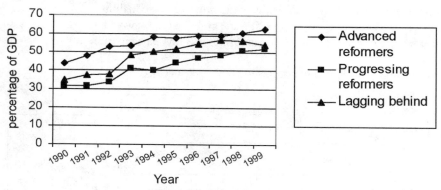

Their participation was, on average, 50 percent lower than their share in Global Gross Product. In general, products and services provided in centrally planned economies were not able to compete successfully on the world market due to low quality and trade barriers.

The following two indicators were chosen to measure the degree to which CEE countries have opened their borders and participated on the world market: Exports of Goods and Services as a Percent of GNP, and Imports of Goods and Services as a Percent of GDP. The first measure, exports, assesses progress toward achieving a competitive position of goods and services on world markets. The second indicator, imports, measures the size of imports compared to the country's own domestic production, providing a measure of wealth and integration with the global market.

Figure 2.11 depicts the export trends in the CEE4 and non-CEE4 countries. At the end of the 1999, the non-CEE4 countries, which had previously lagged behind, surpassed the CEE4. Exports as a share of GDP reached 44 percent among non-CEE4 countries, compared to only 42 percent among the CEE4.

The NIS surpassed the CEE4 in export share of GDP. Does that mean that the non-CEE4 countries had become better competitors than the CEE4? The

Figure 2.10 Contribution of Services to GDP—Early Privatizers

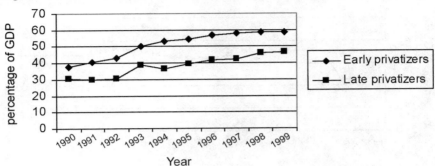

Figure 2.11 Contribution of Exports of Goods and Services to GDP

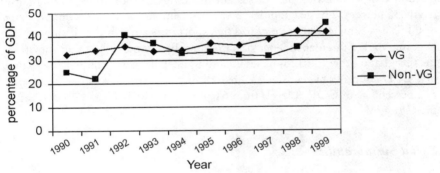

answer lies in the origins of the newly independent states, particularly the Commonwealth of Independent States (CIS), which covers all NIS except the three Baltic countries. These countries, established after the collapse of the Soviet Union, are highly interdependent due to long-term Soviet investment policies that imposed tight technological ties on these former Soviet republics. Therefore, roughly 36–50 percent of all NIS exports went to other CIS countries and Russia. Only about 25–35 percent of the NIS exports were traded in highly competitive markets. In the case of Russia, exports to other NIS states were about 49 percent of all Russian exports in 1999.[14] The exports of the three Baltic countries reached 65 percent of their GNPs, a level comparable to developed countries'. The CEE4 were the most competitive; their exports on the global market reached almost 75 percent of their GNP. Most of these successful countries have sustained this high percentage over the time period analyzed.

Turning to imports, the CEE4 experienced stable annual increases, rising from 29 percent to 46 percent of GDP over the ten-year period studied (Figure 2.12). Among non-CEE4 countries there was a sudden increase from 22 percent in 1992 to 42 percent in 1993, primarily as a result of the collapse of the Soviet Union. (This was similar to the export experience; see Figure 2.11.) However,

Figure 2.12 Contribution of Imports of Goods and Services to GDP

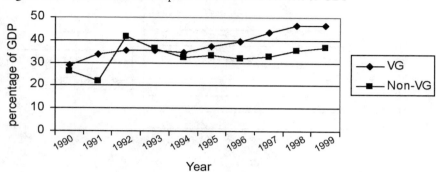

the non-CEE4 countries did not sustain this level but soon dropped to a fairly steady 35 percent. The geographical breakdown among countries in import share of GDP was similar to that described by export indicators in that it showed the ongoing market interdependence of CIS states. By contrast, the trend shown by the CEE4 and other CEE countries including the Baltics was comparable to that of OECD countries. These states enjoyed a high and sustained share of imports as a percentage of GDP, at levels from 61 percent in the Baltics to 73 percent in the CEE4.[15]

Social Sustainability

Economic indicators are not enough to fully assess the sustainability of development. Social welfare indicators are crucial to understanding not only the quality of life but also the sustainability of development achievements in human terms. The three fundamental indicators of social welfare that we will examine are life expectancy, fertility, and infant mortality.

Life Expectancy. Analysis of the relationship between life expectancy and three transition indicators (liberalization, stabilization, and financial institution reform) indicates that success with all three indicators positively affected life expectancy in the CEE countries. Figure 2.13 illustrates that for CEE4 countries, life expectancy rose from 70.8 years (1990) to 73.0 years (1999) and is continuing to trend upward. Similar results were experienced by early liberalizers, early stabilizers, and advanced reformers. By comparison, late liberalizers, late stabilizers, progressing reformers, and those lagging behind experienced losses of 2–3 years in average life expectancy during the same period.

In the case of privatization category 1 (large enterprises), both early privatizers and those lagging behind experienced declines in life expectancy, yet losses for early privatizers were larger than those for late privatizers. Unlike countries in category 1, early privatizers in category 2 (small and medium enterprises) sustained an average life expectancy of 70 years for most of the ten-year period. Late

Figure 2.13 Life Expectancy at Birth

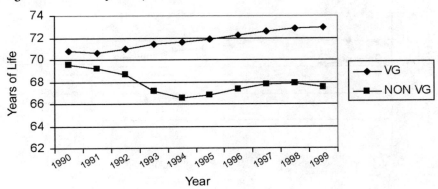

privatizers experienced steadily declining life expectancy, falling from 60 to 58 years.

Fertility. Overpopulation represents a substantial threat to global sustainability. According to the theory of demographic transition, as countries achieve higher incomes and greater educational and employment opportunities, fertility per capita will fall. Thus, falling per capita fertility rates should correlate with higher incomes and greater investment in social capital. Our analysis confirms this outcome in CEE4 as well as among the leaders on all four major transition indicators evaluated in this paper. Figure 2.14 shows similar declines in fertility rates in both CEE4 and non-CEE4 countries.

Infant Mortality. One of the most important social indicators is the infant mortality rate. It is a measure of both quality of life and social sustainability. In our analysis we observed significantly falling infant mortality among CEE4 countries. Figure 2.15 shows how infant mortality rates declined in the CEE4 from seventeen deaths per 1,000 live births in 1990 to eight deaths per 1,000

Figure 2.14 Fertility Rate

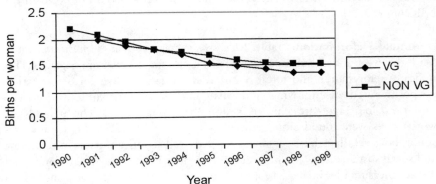

Figure 2.15 Infant Mortality Rate

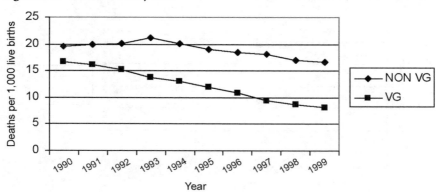

live births in 1999. The non-CEE4 countries' infant mortality rates changed little prior to 1995, first rising slightly then falling back to the 1990 level. Finally, after ten years of transition, the non-CEE4 countries' infant mortality rates fell to seventeen deaths per 1,000 live births. This is the level at which the CEE4 began the decade. Thus, while the CEE4 experienced steadily declining infant mortality through the nineties, the non-CEE4 countries' infant mortality wavered and achieved only minor improvement during the latter half of the decade.

Environmental Sustainability

There are many measures of environmental sustainability, environmental pressure, environmental state or quality, and environmental response. We have chosen to focus on a small subset,[16] due to the lack of consistent data across time in most of the twenty-five countries. Our research uses five major measures: average annual deforestation rate and industrial emissions of CO_2, CO, NO_2, and SO_2. We are aware that the CEE countries and particularly the CEE4 made significant progress in all of the Pressure-State-Response (PSR) areas, including the achievement of significant reductions in industrial air and water emissions during the last decade. Unfortunately, the same improvement did not occur in the NIS countries.

Annual Deforestation. Table 2.1 shows average annual deforestation rates. With the exception of only a few countries, deforestation rates are negative. This indicates reforestation and forest maturation. Negative values for deforestation are positive signs of improving biodiversity in these countries in transition. The lowest (i.e., best) rates are achieved by non-liberalizers (–0.95). The highest (i.e., worst) rates were found among the early liberalizers (–0.27). Taking into account, however, that the early liberalizers achieved significant gains in economic and social sustainability, the limited improvement in the state of environmental quality measured by increased forested land should be regarded as a success. In

Table 2.1 Average Annual Deforestation 1990–2000

Early Liberalizers	%	Late Liberalizers	%
Czech Republic	0	Albania	0.8
Estonia	−0.6	Armenia	−1.3
Hungary	−0.4	Bulgaria	−0.6
Poland	−0.1	Croatia	−0.1
Slovak Republic	−0.3	Georgia	0
Slovenia	−0.2	Kazakhstan	−2.2
GROUP AVERAGE	−0.27	Kyrgyzstan	−2.6
		Latvia	−0.4
Non-Liberalizers		Lithuania	−0.2
Azerbaijan	−1.3	FYR Macedonia	0
Belarus	−3.2	Moldova	−0.2
Tajikistan	−0.5	Romania	−0.2
Turkmenistan	0	Russian Federation	0
Ukraine	−0.3	GROUP AVERAGE	−0.54
Uzbekistan	−0.2		
GROUP AVERAGE	−0.92		

Source: World Bank 2001.

other words, the early liberalizers attained significant economic and social improvements while still seeing some, albeit small, improvements in the state of their environmental health and biodiversity. This finding directly contradicts the claim by some environmental NGOs and others who assert that transition to market economy is sacrificing the ecological well being of CEE countries and NIS.

Emissions of Industrial Pollutants. The level of CO_2 emissions remained almost constant for the early liberalizers over the nine-year period (Figure 2.16) while their GNP per capita increased significantly (see Figure 2.2). This implies a significant structural change toward lower emissions per dollar of GNP, possibly due to an efficiency gain achieved mostly in the use of energy. In the meantime, non-liberalizers reduced their level of pollution but also decreased their

Figure 2.16 Carbon Dioxide Emissions

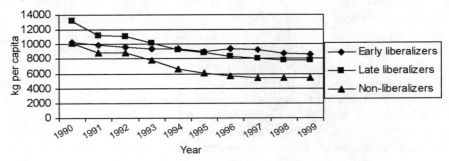

GNP per capita. Their percentage decrease in both CO_2 levels and GNP per capita was approximately 54 percent. As a result, their emissions per dollar of GNP remained about the same. Trends in the other three industrial emissions are similar (although not identical).

Results of Trends Analysis

This examination of quantitative data supports our hypothesis that early liberalization, early stabilization, advanced financial institutional reform, and early privatization contributed significantly to a much smoother, more successful, and more sustainable transition in the CEE4, CEE, and NIS countries. The results of this analysis show clearly that a faster transition appears to be a significant positive contributor to economic, social, and environmental sustainability. We have measured that speed by examining four broad indicators: market liberalization, price stabilization, privatization, and financial institution reform. The countries attaining a faster and more thorough economic transition simultaneously achieved greater sustainability by all of the measures we considered.

A Deeper Look: The Environmental Kuznets Curve (EKC)

The promising results of the trends analysis motivated us to conduct additional research to explore more rigorously the relationship between the transition processes and social outcomes using econometric analysis. Given the observed downward trends in pressure indicators presented earlier, can one conclude that the CEE countries have been successful in reducing regional pollution emissions since the transition to market economies occurred? Not without further analysis. These countries endured a severe recession during the transition, and that alone could account for the majority of the observed declines in pollutant emissions. As discussed above, the level of pollutant emissions depends upon a number of critical factors in addition to the decline in output evidenced during the transition, whose contribution to observed declines needs to be identified. We also hope that we will be able to paint a picture of the success of the CEE countries in developing a unique model of transition.

We test for this "policy EKC" using the twenty-five nations in transition in Central and Eastern Europe and the Newly Independent States as a means of determining the extent to which different transition models (speed of liberalization, type of privatization, extent of financial reforms, and stabilization efforts) may result in more sustainable development measured as declines in emissions in the face of increasing per capita incomes.

Support for this hypothesis should not be misinterpreted to imply that economic growth is good for the environment or that the environment will automatically improve solely with economic growth. However, it does contradict the view that greater economic activity inevitably hurts the environment—a view that is largely based on static assumptions about technology, tastes, and environmental investments. It also allows us to test the hypothesis that the speed of transition to market economies may, under some conditions, be sustainable.

In designing a useful econometric model, we chose measures similar to those already described in the analysis of sustainability trends. We attempted to identify the most important economic, structural, and policy variables that might affect the environment in the region. By including these measures in our mathematical model, we hoped to be able to isolate the effect of each in the transition of Central and Eastern Europe countries to market economies and democratic societies.

Definition of the Environmental Kuznets Curve (EKC)

The environmental Kuznets curve (EKC) hypothesizes that environmental damage first increases with rising income and then declines. This hypothesized relationship between environmental degradation and income per capita thus takes the form of an inverted U. There are several key factors other than income that can also play a role in environmental emissions. These include technology, composition of the economy (structure), environmental investments, openness, and other policy reforms.[17]

There have been several empirical analyses of the EKC hypothesis with mixed results. Our own earlier analysis of the EKC in CEE countries provided support for the hypothesis. We found evidence that rising incomes in the CEE countries initially contributed to higher levels of emissions of key air quality indicators, but that beyond a specific level of income, a "turning point," higher incomes led to reduced pressure on the environment. In our earlier analysis, these turning points were significantly lower than turning points identified for developed countries. Additionally, we found some evidence that policy reform with regard to the openness of the economy, environmental investments, and the degree of privatization was effective in reducing environmental pressure for some selected pollutants.

The most critical were policy reforms implemented quite effectively in CEE4 that resulted in significant increases (two to four times the 1990 level) in environmental investments.[18] The rapid increases in environmental investments were direct responses by newly (and, for the first time since World War II) democratically elected governments to the state of the environment and the public perception of the threat posed by pollution and seriously damaged ecosystems. Reestablishment of democratic rules in CEE countries, including the abolition of state censorship, resulted in a massive wave of publications about the enormous scale of environmental disaster and its ecological and health consequences for the whole region.[19] Several public opinion polls, including Gallup, conducted in CEE at the beginning of the 1990s, clearly indicated that the environment was a top (number one or two) priority in Hungary, Poland, and Russia.[20] So, the governments had to react and, at least in the case of CEE4, introduced significant policy changes.

Serious budgetary limitations and thus enormous pressure for the highest efficiency prompted the governments in CEE4 to revise the scope and enforceability of regulatory instruments and introduce a large variety of economic instruments. These were largely market-based mechanisms (MBM). Thus, through pollution and user fees and product charges, significant public funds were created for environmental protection, and a relatively large (about USD 12 billion) and dynamic environmental market emerged at the beginning of the twenty-first century.[21] The size of this market and its dynamism much better explain the demand for environmental goods and services than does an abstract analysis of changes in income per capita. Contrary to the conventional assumption related to EKC, in CEE this demand was not a result of the appearance of luxury goods and services provided by the environment, but rather a direct reaction to environmental hazards and threats to societal health and life.[22] No other industrial countries have experienced such intensity of pollution and environmental degradation at such low levels of income.[23] For that reason, the classical assumptions of the EKC should be revised and adjusted to particular conditions of the country under consideration.

Another investigation, conducted by Zylicz's team in the Baltic Sea countries in the mid 1990s, showed that there was not a clear indication between the level of income and the willingness-to-pay (WTP) for cleaning up the sea from eutrophication.[24] The rates that relatively low income Poles offered to pay for a clean Baltic were not much different from those offered by Germans with almost three times higher GDP per capita. The CEE4 experience with environmental reforms and creative application of MBM thus presents an interesting lesson for the theory of environmental economics as well as for practitioners designing environmental and sustainable development policies. Unfortunately, most of the CIS, including Russia, has chosen a different path of transformation, resulting in negative consequences for the environment and health.[25]

Economic and Policy Factors Critical to Environmental Quality

The critical factors of environmental significance that have emerged since the transition began in the late 1980s are shown in Table 2.2. Indicators were selected to provide information on the changes in critical factors. Table 2.2 shows expected effects on the environment, a definition of the critical factor and the indicator. The effects of each of these factors on pollution emissions must be sorted out in order to determine the contribution of each to the observed reductions in environmental pressure. Using statistical methods, we can identify how each of the critical factors contributes to observed changes in pollution emissions and make some preliminary determination of the significance and size of these separate effects.

Consumption Patterns. A key objective of the EKC analysis is to examine consumption patterns, as measured by income per capita, to determine the extent

Table 2.2 Factors of Environmental Significance Following Transition *

Factor	Effect on the Environment	Definition	Indicator
Consumption patterns	Rising incomes expected to have a negative effect (increase pressure)	Per capita income; income growth rate	GDP per capita; Δ GDP in 1995 USD
Structure of the economy	Faster privatization expected to have a positive effect (reduce pressure)	Contribution of private sector to the economy	EBRD index
	Manufacturing expected to have a negative effect in most cases (increase pressure)	Contribution of manufacturing sector to the economy	Ratio of value added from manufacturing sector to total GDP
Regulatory structure	Advanced reform expected to have a positive effect (reduce pressure)	Progress in reforming financial institutions	EBRD index
Economic stability	Full stabilizers expected to have a positive effect (reduce pressure)	Efficiency in reducing the rate of inflation	EBRD index
Openness	Full liberalization expected to have a positive effect (reduce pressure)	Complete price liberalization; full current convertibility; almost complete small-scale privatization	EBRD index
	No a priori expectations about the effects of foreign direct investment	Contribution of foreign direct investment to the economy	Ratio of foreign direct investment to GDP
Country-specific effects	Larger countries expected to have a negative effect (increase pressure)	Country size	Population in millions
	CEE countries expected to have a positive effect (decrease pressure)	Individual CEE countries	Country-specific dummy variable
Efficiency gains	Efficiency gains expected to have positive effect (reduce pressure)	Change over time	Time trend

* For the purpose of our analysis we have used the CEE4 as a proxy category for early liberalizers, early stabilizers, early privatizers, and advanced reformers.

to which rising per capita incomes can be expected to contribute to pollutant emission levels and to test the inverse U-shaped relationship. If rising incomes due to transition result in environmental degradation, the claim of sustainable development cannot be made. If, on the other hand, positive income growth and progress towards market reforms lead to a significant reduction in environmental pressure (mostly pollution) and usually to an improvement in environmental quality, development would be considered to be sustainable.

Increased consumption that results from rising incomes is generally expected to increase environmental pressures, *ceteris paribus*. We expect to find that declining incomes have contributed to a reduction in observed environmental pressure. As incomes begin to rise again with economic recovery, an increase in pollutant emissions is expected to follow. Incomes are expected to rise well into the middle of the next century, although debate continues as to whether and at what point rising incomes might begin to slow pollution increases.

The measure of income adopted is gross domestic product (GDP) per capita measured at constant prices in local currency units and expressed in U.S. dollars using the World Bank 1998 conversion factor.[26] It was necessary to use World Bank GDP data to obtain a sufficient time series for all twenty-five countries. To test for inverse U-shaped patterns, income per capita is included in quadratic form.

In order to isolate and measure the effect of income on environmental change, our econometric equation must include measures of all other critical economic and policy factors believed to affect the environment. A discussion of these other indicators, most importantly changes in economic and regulatory structure and openness to international trade, follows.

Structure of the Economy. Because we are testing the hypothesis that economic reforms have not had a deleterious impact on the economy, measures of structural change are included. Changes in the structure of the economy include the development of a private sector and the decline in the manufacturing sector. Both are critical factors predicted to both reduce and increase environmental pressures. The changing industrial structure reduced inefficient heavy industry, which could explain some portion of the observed downward pressure on air quality indicators.

The degree to which market reforms have been achieved is also expected to have a significant effect on pollutant levels, although there is debate about the observed direction of its effect. Some argue that privatization and industrial restructuring have been promoted with little awareness of the potentially negative environmental impacts.[27] Others believe that market reform has a positive effect, at least through gains in efficiency of resource use, if nothing else.

Also important in the transition process in these countries is the role of Foreign Direct Investment (FDI). Here, it is initially measured as net inflows of foreign direct investment as a percent of GDP. There are no *a priori* expectations about the direction of the effects of FDI on emissions levels, although for sustainable development it is hoped that investments would be made to reduce pressure on the environment. Manufacturing value added (measured as a percent of GDP) is used to capture the changes in the structure of the economy resulting from transition.

Regulatory Structure. The forces of economic transformation alone are not likely to reduce environmental pollution. Regulatory structures and investments are also needed to adjust for externalities and provide more environmentally friendly technologies. Environmental regulation should reduce pressure on the environment. With appropriate regulatory institutions, it is possible that economic growth can continue without undue harm to the environment. Environmental expenditures (percent of GDP) are used as a proxy for regulatory intensity. This captures investments in environmental improvement as well. Unfortunately, this indicator was only available for CEE4, the Baltics, and a few other countries. For most of the NIS, reliable data were not available regarding either regulations or environmental expenditures.

Other factors. Population and time are included to account for size effects as well as changes not related to those captured by the critical factors.

Measuring the Impact on Environmental Quality

The initial empirical analysis of the effects on economic growth and transition on the environment focuses on air quality in the region. While it is preferable to rely on indicators of environmental quality that measure the state of the environment directly, such data are not consistently available for these countries. Typically, pressure indicators are relied upon as a proxy for environmental quality in such cases. Changes in pressure indicators imply changes in environmental quality. The four dependent variables, i.e., the key indicators measuring air pollution pressures, are as defined in Table 2.3.

A regional analysis is critical, given that these key air pollutants have global as well as local effects. Furthermore, the indicators developed for the Pressure-State-Response (PSR) system are most appropriate to regional analysis.[28]

Model Specification

An econometric analysis is conducted using a pooled-time series data set for the twenty-five countries covering the period 1993–99. Earlier data are not reliable. In accordance with Kuznets's hypothesis, we designed a reduced form of the general linear fixed effects regression model, and also made appropriate changes for the various specifications defined in Table 2.4. The reduced form assumes that environmental outcomes are related to predetermined endowments and economic measures. Clearly, some interdependence in environmental indicators is probable.

Several alternative model specifications were estimated (see Table 2.4) using the EKC policy augmented framework. The initial model (Model 1) applied to

Table 2.3 Key Air Quality Pressure Indicators

Pollutant	Indicator definition and unit of measurement	State indicator	Alternative indicator definition	Scope of effect
Emissions of carbon dioxide (CO_2)	Total (or net) emissions of CO_2 in kt per year	CO_2 concentrations, global temperatures	None	Global climate change
Emissions of carbon monoxide (CO)	Total emissions of CO in kt per year	CO concentrations	Emissions of CO per unit of area; emissions per capita	Global, regional climate change
Emissions of sulfur dioxide (SO_2)	Total emissions of SO_2 in kt per year	Concentration of sulfates (SO_2), global temperature	An aggregated indicator for aerosol particles	Global, regional and local climate change; acidification
Emissions of nitrogen oxides (NO_2)	Total anthropogenic emissions of NO_2 in kt per year	Concentrations of atmospheric nitrous oxide; global temperature	Emissions of NO_2 per unit of area; aggregate indicator for CO_2, CH_4, and NO_2 emissions based on global warming potential	Global and regional climate change; ozone depletion

Source: Eurostat Methodology Sheets, 1998

all four pollutants (CO_2, CO, SO_2 and NO_2) tested for effects on emissions of income, financial reforms, openness, structure of the economy, and country size. Model 2 added country-specific effects for the CEE4 as an example of those countries that had the speediest transition (in terms of liberalization, stabilization, privatization, and reform). Model 3 added time effects.

Results of Econometric Analysis

Preliminary econometric evidence supports the trends analysis presented earlier. The power of the economic and policy variables to predict changes in the air quality indicators was quite strong; adjusted R2 for each model ranged from .79 to .96. Signs—i.e., direction of impact on the environment—were generally as expected. Incomes at first increased emissions pressure and then reduced it. For three out of four pollutants these estimated parameters were statistically different than zero.

Table 2.4 Econometric Results

Pollutant: CO_2	Model 1	Model 2	Model 3	Pollutant: CO	Model 1	Model 2	Model 3
Variables:				Variables:			
Time	$-^*$	$-^*$		Time	$-^{**}$	$-^{**}$	
GDP	$+^*$	$+^{**}$	$+^{**}$	GDP	$+$	$+^{**}$	$+$
GDP^2	$-$	$-$	$-$	GDP^2	$-$	$-^{**}$	$-^{**}$
$BKREF^a$	$-$	$-$	$-$	$BKREF^a$	$+^{**}$	$+$	$+$
FDI	$+$	$+$	$+$	FDI	$-$	$+$	$-$
MFD^b	$-$	$-$	$-$	MFD^b	$+^{**}$	$+^{**}$	$+^{**}$
POP^c	$+^*$	$+^*$	$+^*$	POP^c	$+^*$	$+^*$	$+^*$
Czech Republic		$+$	$+$	Czech Republic		$-$	$-$
Hungary		$-$	$-$	Hungary		$-$	$-$
Poland		$-^{**}$	$-^{**}$	Poland		$+^*$	$+^*$
Slovakia		$+$	$+$	Slovakia		$-$	$-$
1993			$+$	1993			$+^{**}$
1994			$+$	1994			$+$
1995			$+$	1995			$+$
1996			$+$	1996			$-$
1998			$-$	1998			$-$
1999			$-$	1999			$-$
Adjusted R^2	.96	.96	.96	Adjusted R^2	.90	.95	.95

Pollutant: SO_2	Model 1	Model 2	Model 3	Pollutant: NO_2	Model 1	Model 2	Model 3
Variables:				Variables:			
Time	$-^*$	$-^*$		Time	$-^*$	$-^*$	
GDP	$+$	$+$	$+$	GDP	$+^*$	$+^*$	$+^*$
GDP^2	$-$	$-$	$-$	GDP^2	$-^*$	$-^*$	$-^{**}$
$BKREF^a$	$+^{**}$	$+$	$+$	$BKREF^a$	$+$	$+$	$+$
FDI	$-$	$-$	$-$	FDI	$+$	$+$	$+$
MFD^b	$+^*$	$+^{**}$	$+^{**}$	MFD^b	$+^{**}$	$+$	$+$
POP^c	$+^*$	$+^*$	$+^*$	POP^c	$+^*$	$+^*$	$+^*$
Czech Republic		$+^*$	$+^{**}$	Czech Republic		$+^*$	$+^*$
Hungary		$+$	$+$	Hungary		$-$	$-$
Poland		$+^*$	$+^*$	Poland		$+^*$	$+^*$
Slovakia		$-$	$-$	Slovakia		$+$	$+$
1993			$+^{**}$	1993			$+^*$
1994			$+$	1994			$+$
1995			$+$	1995			$+$
1996			$+$	1996			$+$
1998			$-$	1998			$-$
1999			$-$	1999			$-$
Adjusted R^2	.69	.79	.79	Adjusted R^2	.94	.96	.96

* Significant at .01 level
** Significant at .05 level
[a] BKREF is an indicator of financial institutions' (mostly banks) reforms
[b] MFD is an indicator of changes in the national economy structure (GDP produced in manufacturing industries)
[c] POP is an indicator of a country's size measured by population

The effects of other policy variables were not as consistent, although country size always had a negative effect on sustainability (indicating increased pressure). Preliminary evidence is consistent with expectations. The time indicators provide evidence that the region is still on this path, as the three years since 1999 confirm the downward trend. Evidence of efficiency gains in the CEE4 is obvious with regard to trends in CO_2 and SO_2 emissions. In the CEE4, particularly Poland, rising incomes have contributed to rising NO_2 as demand for automobiles has increased.

Results for these models were quite robust with respect to the turning points derived as shown in Table 2.5. For CO_2, the range of income at which emissions begin to decline is between USD 7,148 and USD 7,561; for CO the range is USD 3,701 to USD 4,418; for SO_2 the range is USD 2,172 to USD 3,439 and for NO_2 it is USD 5,722 to USD 6,364. These turning points are below those observed in developed countries. For example, in developed countries, the average for CO_2 is approximately USD 8,000.

Conclusions

Transition does not necessarily affect sustainability adversely as Stiglitz argued; at least, this has not happened in the Central European model of transition. Any conclusions regarding transition processes must include the Central European model experiences in the CEE4 as well as the Baltic States. Even including all twenty-five countries, there is evidence in support of the EKC hypothesis.

This initial analysis of the available data for the CEE countries and NIS provides some important information regarding the relationships between key pollutant emission indicators and critical factors related to the economy and institutional reforms that have occurred in the CEE countries over the past decade. Wise policy choices clearly are needed to speed up and continue the observed trends; they are critical to future gains in environmental quality. Nonetheless, in the absence of further policy initiatives to assure environmentally sound restructuring, future increases in income in the near term can be expected to increase pressure on the environment.

This analysis also provides preliminary evidence that the demand for environmental quality is likely to occur for most pollutants at lower income levels than previously believed. This has important consequences for estimates made by policy makers about the growth of emissions in the future and the costs of achieving environmental gains.

Table 2.5 Income Levels at Emissions Turning Points

	CO_2	CO	SO_2	NO_2
Model 1	USD 7,561	USD 3,701	USD 3,439	USD 5,722
Model 2	USD 7,148	USD 4,456	USD 2,304	USD 6,385
Model 3	USD 7,210	USD 4,418	USD 2,172	USD 6,364

Any conclusions must necessarily remain tentative, given the simple econometric models employed in this analysis and the preliminary nature of the sustainability indicators available to us. More work needs to be done to develop better measures of trade and foreign direct investments. While the results for air quality are encouraging, there is some indication that a coordinated regional strategy for reducing air pollution is needed. Countries should also pay closer attention to foreign direct investments to assure that they reduce environmental pressure. This is important given that foreign investments are expected to comprise a large share of total investment in the near future.[29]

Notes

1. See Sandra Archibald and Zbigniew Bochniarz, "Environmental Outcomes Assessment: Using Sustainability Indicators for Central and Eastern Europe to Estimate Effects of Transition on the Environment," in *World Congress of Environmental and Resource Economics: Book of Abstracts* (Venice: AERE & EAERE, 1998), 320 and full publication on the website; and Sandra Archibald and Zbigniew Bochniarz, "An empirical analysis of the environmental Kuznets curve in Central and Eastern European transitional economies," in *Ninth Annual Conference of the European Association of Environmental and Resource Economics: Book of Abstracts* (Oslo: EAERE, 1999), 36 and full publication on the website.
2. Albania, Armenia, Azerbaijan, Belarus, Bulgaria, Croatia, Czech Republic, Estonia, Georgia, Hungary, Kazakhstan, Kyrgyzstan, Latvia, Lithuania, FYR Macedonia, Moldova, Poland, Romania, Russian Federation, Slovak Republic, Slovenia, Tajikistan, Turkmenistan, Ukraine, and Uzbekistan.
3. See Joseph Stiglitz and David Ellerman, "Not Poles Apart: 'Whither Reform?' and 'Whence Reform?'" *Journal of Policy Reform* 4 (2001): 325–39; and Joseph Stiglitz, "Quis Custodiet Ipsos Custodies? Corporate Governance Failures in the Transition," keynote address at the *Annual Bank Conference on Development Economics* (Paris: World Bank, 1999).
4. See Marek Dabrowski, Stanislaw Gomulka, and Jacek Rostowski, "Whence Reform? A Critique of the Stiglitz Perspective," *Journal of Policy Reform* 4 (2001): 291–324.
5. Stiglitz, "Whither Reform?"
6. See Masahiko Gemma, "Industrial Development in Transition Economies: Lessons and Implications," *Waseda Studies in Social Science* 1 (2000): 19–31.
7. See Zbigniew Bochniarz, *The Research Methodology of the Project on Assessing Environmental and Social Sustainability of Transitional Economies,* JSPS sponsored project at University of Minnesota and Waseda University, October 2000.
8. European Bank for Reconstruction and Development, *Transition Report 1999* (London: EBRD Publication Desk, 1999), 64.
9. See Dabrowski, Gomulka, and Rostowski, "Whence Reform?" 313.
10. EBRD, *Transition Report 1999,* 64.
11. EBRD, *Transition Report 1999,* 64.
12. David Pearce, *Blueprint 3: Measuring Sustainable Development* (London: Earthscan, 1993), 28.
13. The annual values for each group of countries were obtained by weighting the value for each individual country according to the proportion of its population size (or GDP, where necessary) in the cumulated population of the whole group (or cumulated GDP respectively). This method was used in order to take into account the differences in population (or GDP) size between the countries.

14. United Nations Economic Commission for Europe, *Economic Survey of Europe* (Geneva: UN Publications, 2001), 143.
15. Ibid.
16. See Bochniarz, *Research Methodology*, 3.
17. See David I. Stern, Michael S. Common, and Edward B. Barbier, "Economic Growth and Environmental Degradation: The Environmental Kuznets Curve and Sustainable Development," *World Development* 24 (1996): 1151–60.
18. OECD, *Environment in the Transition to a Market Economy: Progress in Central and Eastern Europe and the New Independent States* (Paris: OECD, 1999).
19. Maria Guminska and Andrzej Delorme, eds., *Kleska Ekologiczna Krakowa* (Cracow: PKE Krakow, 1990); Zbigniew Bochniarz, Expert Testimony to the U.S. Congress, in *U.S. Environmental Initiatives in Eastern Europe: Hearing before Subcommittee on Transportation and Hazardous Materials of the Committee on Energy and Commerce, House of Representatives, 101st Congress, 2nd sess., 23 April 1990*, Serial No. 101-140 (Washington: U.S.G.P.O., 1990); Barry S. Levy, ed., *Air Pollution in Central and Eastern Europe: Health and public Policy* (Boston: Management Sciences for Health, 1991).
20. "Environment a Top Concern, Poll Reveals" *Budapest Week: Hungary International Weekly* 2, no. 17, (July 2–8, 1992).
21. Patrick Francis, ed., *National Environmental Protection Funds: Case Studies of Bulgaria, the Czech Republic, Hungary, Poland and the Slovak Republic* (Budapest: Regional Environmental Center, 1994); OECD, *Environment in the Transition to a Market Economy*.
22. Bochniarz, Expert Testimony; Levy, *Air Pollution in Central and Eastern Europe*.
23. Zbigniew Bochniarz, *In Our Hands; United Nations Earth Summit '92: Capacities and Deficiencies for Implementing Sustainable Development* (Geneva and Minneapolis: UNDP, 1992).
24. Agnieszka Markowska and Tomasz Zylicz, "Costing on International Public Good: The Case of the Baltic Sea," *Ecological Economics* 30 (1999): 301–16.
25. OECD, *Environment in the Transition to a Market Economy*.
26. Zbigniew Bochniarz and David Toft, "Free Trade and the Environment in Central Europe," *European Environment* 5 (1995): 52–7.
27. See David W. Pearce and Jeremy J. Warford, *World Without End: Economics, Environment and Sustainable Development* (New York: Oxford University Press, 1993).
28. See European Official Statistics, *Eurostat* (Luxembourg: Office for Official Publications of the European Communities, 1997).
29. See Zbigniew Bochniarz, Wladyslaw Jermakowicz, and David Toft, "Strategic Foreign Investors and the Environment in Central and Eastern Europe," in *Innovation, Technology and Information Management for Global Development and Competitiveness*, ed. Erdener Kaynak and Tunc Erem (Hummelstown, PA: International Management Development Association, 1995).

Chapter 3

SUSTAINABILITY AND EU ACCESSION

Capacity Development and Environmental Reform
in Central and Eastern Europe

Stacy D. VanDeveer and JoAnn Carmin

Central and East European states enacted and implemented sweeping changes as they prepared to gain membership in the European Union. As a result of severe degradation, environmental protection and remediation were important political priorities for these countries, particularly in the early phases of transition away from Soviet-style rule. In the fifteen-plus years since the fall of communism, CEE policy makers have developed new institutions and formulated, enacted, and implemented new policies for environmental protection.[1] Furthermore, some significant reductions in pollution have been achieved.[2] While the eastward enlargement of the EU promises to promote greater environmental policy coordination and standardization across Europe, numerous institutional and environmental challenges remain.

As a means for understanding sustainability in the "new" Central Europe, this chapter examines capacity development, looking at different sources of aid and how they shaped domestic policy and institutions in the period leading up to accession.[3] General sources of international support since 1989 are reviewed, with particular attention given to the implications of EU enlargement for environmental policy institutions and civil society actors within the region. We maintain that, although there were numerous international sources of capacity development in the initial years after the fall of the communist regimes, the EU consolidated its influence in the 1990s and, as a result, has had the greatest impact on CEE environmental policy and institutions. Further, we suggest that the common assumption of unidirectional diffusion of environmental policy institutions and information from West to East will likely require revisiting, given that it ultimately limits both CEE and EU potential for achieving innovations in environmental policy institutions and implementation.

Environmental Change and Capacity Development in CEE

The transition from communism to democracy was expected to bring sweeping reforms across the social, political, and economic arenas. With the heightened attention given to environmental concerns and the role that these issues played in the overthrow of the regimes, many officials and civil society leaders anticipated that CEE countries would demonstrate strong environmental leadership.[4] While this potential has not been fully realized, the expansion of environmental policy and law in the region has been dramatic.[5] For example, Czechoslovakia formed the Ministry of the Environment in 1990 and enacted a comprehensive body of environmental laws in 1992. Poland strengthened the State Inspectorate for the Environment, putting it under the auspices of the existing Ministry of the Environment, and updated its Nature Protection Act in 1991. These developments reach beyond CEE countries in the first wave of accession. Romania, for instance, though it has much environmental policy left to harmonize with that of the EU, has made extensive changes in environmental law and regulation.[6]

Changes have taken place not only in the institutional arena, but also with respect to environmental quality. Areas of severe degradation, from the Balkans to the Baltic States, resulted from practices prevalent during the communist era. Since 1989, there have been numerous instances of successful and ongoing efforts at remediation of past pollution and prevention of further degradation. In Poland, major polluting plants were closed, in the Czech Republic they were ordered to cut emissions, and in the former GDR, the use of brown coal has been significantly reduced. These and other changes have contributed to improvements in air quality, including reductions in solid, NO_2, hydrocarbon, and SO_2 emissions.[7] Similarly, newly implemented practices and policies have led to improvements in the quality of wastewater discharges and surface water quality in EU accession countries. Additionally, across CEE countries, wastewater treatment capacity has been expanded and emissions of many hazardous substances have declined.[8]

Capacity Development for the Environment (CDE)

The dramatic expansion of the scope of environmental institutions and stringency of environmental policy seen in many CEE countries, as well as some improvements in environmental protection and remediation, are the result of international support and capacity building programs. "Capacity building," though widely alluded to in international organizations, assistance programs, and scholarship, often has no clear definition and does not refer to a common set of strategies among its users.[9] As used here, the term refers to efforts and strategies intended to increase the "efficiency, effectiveness, and responsiveness of government performance."[10] As such, capacity building is a general term, encompassing capacity enhancing, strengthening, and development of any number of "sites" including government bodies, NGOs and civil society, independent unions, political parties, scientific and technical communities, private sector actors, and publics. Furthermore, various agents can seek to build capacity across borders, including states, multilateral organizations, NGOs, firms, and individuals.

Though scholarly definitions may differ, discussions among various donor agencies have begun to converge on a complex and nuanced perspective regarding capacity building. There is now a recognition that it is important to consider capacity at different levels ranging from the individual to the institutional, that a multiplicity of state and nonstate organizations (public, private, and civil society actors) contribute to national capacity, and that it is important to empower and include many relevant actors in decision-making processes (i.e., it is as important to utilize capacity as it is to build it). At the same time, the relationship between technical assistance and capacity building has also been reexamined, with the realization that specific efforts are required for technical assistance activities to build capacity (e.g., through strengthening of institutions and human resources). Activities targeted purely at assisting countries in the completion of specific tasks or moving toward short-term goals may not only fail to build local capacity, but may actually be counter productive in that regard. In fact, attention to process as well as outcomes has led "capacity development" slowly to supersede "capacity building" as the term of choice, with the understanding that the former is a process- as well as an outcome-based approach.[11]

The "Rush" to Provide Support

In the years immediately following the fall of the regimes, numerous interstate and transnational actors attempted to aid the transition to democracy through traditional capacity building approaches. With stories of severe environmental degradation receiving significant play in the press, many Western countries and donors rushed to provide financial support for pollution remediation and institutional development.[12] Between 1990 and 1996, G-24 countries committed almost USD 81 billion in grant aid to the region to promote economic restructuring, democracy building, and general quality of life programs.[13] According to Kolk and van der Weij (1998), approximately ECU 3.5 billion was invested in the environment by international governments from 1990 to 1995.[14] Most funds invested in environmental remediation and protection in the initial years of the transition came in the form of loans from multilateral development banks.[15] Grant-making programs were dominated by bilateral assistance efforts by states across Western Europe as well as the United States and Japan. Germany and Denmark were the largest European donors, dedicating approximately ECU 392 and 118 million respectively to the environment. The United States committed approximately 231 million ECU to environmental issues during this same period, with much of the funding allocated through the Support for East European Democracy Act (SEED) passed in 1989.[16] Intergovernmental donors such as the World Bank also focused on environmental issues, such as building environmental economics capabilities or strengthening regulatory mechanisms and institutions for improved environmental management through their cross-cutting capacity development programs.

Capacity building efforts were not limited to financial support: states, national and multilateral governmental agencies, private foundations, NGOs, and private firms from around the world offered to provide also scientific, technical, and policy guidance.[17] The Environment for Europe (EfE) process is an example of a networking and information sharing effort across continental Europe that was influential in CEE. The initial EfE meeting, held in the Czech Republic in 1991, brought together environmental ministers from across Europe with the goal of coordinating policies for environmental protection. Building on this foundation, a second meeting held in 1993 established the Environmental Action Program Task Force (EAP-TF) to build institutional capacity, sponsor environmental assistance and investment programs, and assist national and local governments in applying administrative techniques that would help integrate the environment into existing reformation processes.[18] The EfE process offers opportunities for environmental officials to set joint goals and priorities for research, policy making, and information sharing across the continent. Western European officials' commitment to such cooperation helps to explain why the European Environment Agency in Copenhagen is the only EU-affiliated body to admit CEE states prior to their membership in the Union itself.

As with funding, technical assistance was not limited to European sources. For example, the SEED Act, along with the Freedom Support Act of 1991, made it possible for the U.S. Environmental Protection Agency (USEPA) to send specialists to the region. Such efforts promoted cooperation and policy collaboration between CEE officials and a diverse set of organizations including USEPA, the World Bank, Resources for the Future, and a host of North American experts in universities and the public and private sectors.[19] Working in collaboration with CEE government agencies, representatives from USEPA provided analysis and assistance as they addressed a wide range of water, air, and waste disposal problems. In addition to working to resolve pressing environmental issues, they promoted the transfer of relevant environmental and information technologies such as the use of Geographic Information Systems, sponsored technical exchanges, and conducted training and education programs.[20]

International and intergovernmental assistance that was channeled to the region after regime change helped many of these countries realize improvements in environmental quality and establish new environmental institutions. While environmental remediation and institutional change were facilitated through direct aid, the speed of these developments was enhanced by familiarity with Western policy and approaches. In the relatively more open Central European countries such as Poland and Hungary, the influence of Western environmental ideas and assistance stemming from exchange programs predates the 1989 collapse of the communist regimes. The significant and lasting effects of these early international influences promoted understanding and receptivity toward Western knowledge and policy instruments.[21] Although it is difficult to directly trace the influence on policy of specific ideas and norms diffused through capacity building programs, previous experience in combination with the presence of experts from North American and Western European institutions during the transition may help explain why these countries readily embraced various market-style approaches to environmental protection. This propensity has been evidenced in the adoption of some innovative environmental financing and market-based environmental policy instruments across the CEE region, including emissions trading schemes, environmental taxes and environmental funds, debt-for-nature swaps, environmental liabilities policies, and incentives for expansion of CEE environmental markets.[22] It must be noted, however, that much EU environmental regulation tends to be command-and-control oriented, likely limiting possible uses for more "market-oriented" policy instruments.

International governments and intergovernmental organizations provided direct support for environmental NGOs and channeled funds to these groups through civil society development and democracy building programs.[23] Some funding from the EU's Poland and Hungary Assistance for Economic Restructuring (PHARE) program went toward environmental education and capacity building of CEE environmental NGOs.[24] One of the most significant developments in this era was the formation of the Regional Environmental Center (REC). REC was created in 1990 by the United States government, the Commission of European Communities, and the Hungarian government with the purpose of building the capacity of environmental NGOs through training, education, and direct support of environmental initiatives.[25] In the early stages of the transition, international NGOs such as the World Wide Fund for Nature and The World Conservation Union, and international foundations such as The Rockefeller Brothers Foundation, the German Marshall Fund, and Milieukontakt Osteuropa supported both environmental projects and CEE environmental organizations. Support from international NGOs also was directed to intermediary funders in the region. REC and the Environmental Partnership for Central and Eastern Europe were major recipients of international foundation and government support, which the organizations then disseminated to regional environmental NGOs with the intent of funding their projects as well as building their administrative and organizational capacities.[26]

EU Influence Consolidated

International aid, equipment, and expertise contributed to the environmental capacity of government institutions and NGOs in the region. Although numerous ideas and practices were diffused from the West, by the mid 1990s the most significant force shaping the environmental policy agendas of CEE states was the desire to join the European Union.[27] A number of states applied for EU membership between 1994 and 1996. In March 1998, accession negotiations were started with Hungary, Poland, Estonia, the Czech Republic, and Slovenia. Subsequently, accession negotiations were opened with Bulgaria, Latvia, Lithuania, Romania, Slovakia, and Slovenia. In 2002, thirteen states were formally negotiating EU membership, including these ten CEE countries and Cyprus, Malta, and Turkey. Accession negotiations were finalized with ten countries in December 2002, with eight CEE (Czech Republic, Estonia, Hungary, Latvia, Lithuania, Poland, Slovenia, and Slovakia) and two Mediterranean (Cyprus and Malta) countries joining the EU in 2004. Accession talks continue with Bulgaria, Romania, and Turkey.[28]

The formal agreements between the EU and candidate states specify the terms of adoption, implementation, and enforcement of the body of EU law, regulations, standards, and norms (the *acquis communautaire*). The *acquis* is organized into thirty-one chapters across a host of thematic areas with the environment constituting one such area (Chapter 22). While negotiations have given the new member states some opportunities to phase in compliance, over the past ten years the majority of domestic environmental legislation and institutions in these countries has been developed to approximate and implement the environmental chapter of the *acquis*.[29] Transposing the environmental chapter of the *acquis* requires community framework legislation, measures on international conventions, reduction of global and transboundary pollution, biodiversity protection measures, and measures ensuring product standards.[30] Negotiations on the environmental chapter of the *acquis* are provisionally closed for all ten of the new CEE accession countries with the Czech Republic, Estonia, Hungary, Poland, and Slovenia having been granted transition periods for urban wastewater treatment and most states receiving additional time to implement directives for the recovery and recycling of packaging waste.[31]

Through the 1990s, countries in the region became increasingly committed to EU membership. While the European Commission continued to support the expansion of environmental expertise across legal, regulatory, scientific, and technical domains as well as funding public education campaigns, foreign assistance from the United State's government and many multilateral organizations and North American foundations declined. By 1995, significant sums had already been dedicated to CEE states for environmental issues by European intergovernmental donors including the EBRD (ECU 667 million), EU (ECU 351 million), and European Investment Bank (ECU 257 million).[32] In general, the EU continued to provide financial and administrative assistance for structural policy (ISPA), agriculture (SAPARD), and administration (PHARE). A wide variety of initiatives have received support, but a large percentage of EU assistance to CEE countries has prioritized projects designed to "harmonize" CEE policy and practices with EU environmental directives and regulations. For example, the PHARE Twinning program targets environmental policies and institutions and LIFE, the

Financial Instrument for the Environment, introduced by the EU in 1992, helps EU member states finance nature conservation and the implementation of Community environmental policy. Although not EU members at the time, Estonia, Hungary, Latvia, Romania, the Slovak Republic, and Slovenia were able to take advantage of this program.

By the latter half of the 1990s, the EU was the dominant source of financial and technical support in the region. The consolidation of EU influence extends beyond the public sector, with a number of programs dedicated to enhancing the organizational and political capacities of CEE civil society actors. PHARE funds were still disseminated directly to environmental NGOs to carry out environmental projects during this period.[33] The Regional Environmental Center continued to offer education programs and serve as an intermediary funder for environmental NGOs in the region. These activities received ongoing support from governments around the globe, including the EU and many individual European countries. In recognition of the need to expand inclusion of NGOs in environmental policy-making, the Sixth Environment Action Programme was approved by the European Parliament in January of 2002 to provide funds to these groups, including those from candidate countries. Further, the Aarhus Convention was adopted in 1998 to ensure access to environmental information and promote civil society participation in environmental decision making.

After the fall of the communist regimes, governments from around the world came to the aid of CEE states, providing funds and technical support to assist in economic restructuring and democracy building. Environmental remediation and protection initially were important agenda items for CEE nations and their supporters. In the early years of transition, the field was crowded with countries from around the globe offering to provide aid. However, at the time that some countries were completing their environmental aid programs, CEE states were becoming more intent on joining the EU and harmonizing their environmental laws, regulations, institutions, and practices with those in the *acquis*. Practices such as environmental impact assessment that were initially developed largely by North American experts have been altered to harmonize with EU expectations. Likewise, EU environmental law and regulation require CEE states to (re)configure regulatory practices such as permitting and environmental monitoring systems, access to information rights and procedures, and enforcement and inspection regimes. Further, while the EU has helped support and sustain numerous environmental NGOs, it also is a limiting factor. In recent years, the agendas of many groups have been oriented toward assessing the merits and limitations of specific aspects of the environmental *acquis,* while their means of activism are now structured by EU laws governing participation.[34] Ideas and norms from North America and elsewhere in the world are still diffused and adopted in these countries. As these examples suggest, however, the continuity of financial and technical support, combined with the desire to accede to the Union, has made the EU the dominant reference point and source of influence for CEE environmental agendas, policies, and practices.

Challenges for Capacity Development for the Environment in CEE

Traditional capacity building programs have led to significant changes across the CEE region in areas such as environmental law, regulation, assessment, evaluation, expertise, and NGO effectiveness.[35] Financial assistance under cofinancing arrangements demonstrates that partial support can get needed and beneficial programs off the ground, helping to improve sewage treatment, drinking water, nuclear power plant safety, air quality, and habitat protection.[36] The transfer of technical expertise and knowledge demonstrates that institutions can be developed and maintained. Further, administrative support and funding have enhanced the presence and activities of environmental organizations. A number of regional organizations have become independent advocates for wide-ranging sustainable development initiatives, while many national organizations are now recognized as having the expertise to make important contributions to international, regional, national, and local environmental decision forums.[37] Although significant changes have taken place as a result of capacity building programs, the EU and CEE countries still face numerous challenges, particularly in the financial and institutional arenas.

Financing Environmental Change

No amount of foreign aid, or EU "structural adjustment funds," will cover the costs being incurred by CEE governments and publics as they seek to adopt and implement the environmental *acquis*. For years to come, CEE officials will have to continue to be creative in their efforts to raise the necessary funds and engender needed behavioral changes. In 1997, the EC estimated that the total cost of adopting the environmental *acquis* would exceed ECU 120 billion.[38] While some reports suggest that this figure is inflated,[39] it provides a reference point for understanding the level of investment required by candidate countries. Accordingly, the Czech Republic invested ECU 1 billion annually on the environment between 1994 and 1999 (or 2.4 percent of GDP) while Poland was spending approximately ECU 2 billion (or 1.7 percent of GDP). Even with these investments, both countries must increase spending to meet their own estimates of the costs of enlargement.[40] Furthermore, much of the international environmental assistance to date was targeted at legal and regulatory harmonization efforts and a set of large (cofinanced) investments. The larger costs of widespread national implementation of the environmental *acquis* are yet to come. In particular, funds for implementation at the local level around wastewater treatment and waste disposal remain extremely scarce. Of course, the costs of implementing the EU environmental *acquis* are only a fraction of the total costs associated with EU membership in CEE countries. Efforts to cover the environmental costs of EU membership—and the cost of other domestically and internationally driven environmental protection measures that will likely be required in the coming years—will compete with the costs of funding other domestic priorities and of implementing other areas of EU law and regulation.

As such, the costly problems faced by CEE officials will likely require continuing policy experimentation and innovation on their part. For example, the scarcity of public sector resources has encouraged CEE policymakers to institute taxation and fees for resource extraction, pollution emissions, water and energy use, packaging, and waste processing.[41] These policies have raised funds for public financing of environmental and non-environmental programs and projects in CEE countries—though the percentage of total environmental fees and taxes dedicated to environmental investments varies substantially. Attention to the need to "get the incentives right" in environmental policy has increased globally in the last decade. Furthermore, well-designed public policy can shape markets in more environmentally sensitive ways, for instance the so called "green taxes" that can shape demand and offer incentives to conserve or increase the efficiency of resource use.

The challenges posed by the costs of implementing many aspects of EU environmental policy, together with the ongoing need in CEE countries for additional foreign investment and economic growth, have the potential to make various market-oriented and taxation-based incentives an attractive option for the region's policymakers. While infrastructure development may require direct investment, particularly at the level of municipalities, in an increasingly integrated European market, market incentives have the potential to reinforce environmental norms and foster the adoption of environmentally sound technologies that can improve the environmental performance of firms and governments.[42] In particular, changes in domestic incentives for various actors may be ushered in as CEE states and firms gain access to European markets and international investments. For instance, pollution emissions from the electricity generation and chemical manufacturing sectors in Poland and Bulgaria suggest that firms reduce emissions in response to incentives offered by European market integration and to domestic pressure.[43] Furthermore, access to information about pollution emissions or environmentally damaging practices may shape public demand and preferences.

Foreign investment appears to accelerate these dynamics in CEE states, because many of the largest foreign investors are from EU member states where relatively high standards are already in place and publics are generally mobilized around environmental issues. Furthermore, some CEE environmental officials admit that they often pursue relatively strict enforcement of environmental laws and regulations among foreign investors because of the general assumption that these investors have "deeper pockets" to pay for environmental compliance.[44] Thus, incentives for domestic CEE environmental officials are also altered by the entry of foreign investors into domestic markets. Lastly, foreign investors can bring with them certain environmental norms and expectations. As Andonova's (2004) research suggests, investors from relatively high environmental standard countries such as Sweden and Germany often apply EU pollution control standards as they upgrade facilities because these are already part of their standard procedures and because they fear additional scrutiny from publics in their home countries.[45] Nevertheless, the significant costs of implementing the environmental *acquis* will pose formidable challenges to national and local CEE authorities for years to come.

Challenges of Institutional Change

Although improvements in state and civil society capacities have been achieved, there are still several important gaps that must be addressed to implement and enforce fully the environmental chapter of the *acquis*.[46] State capacities for ongoing administration, technical monitoring, and enforcement are weak. In many instances, the national environmental ministries are poorly staffed and CEE environmental officials are overwhelmed by the number of international and domestic laws and regulations in their portfolios.[47] These limitations are even more pronounced at the regional and local levels, where limited support for innovation often results in the implementation of readily available solutions rather than the generation or adoption of innovative ones.[48] Many local bodies are tasked with monitoring, enforcing, and implementing a host of complex and expensive environmental policies. However, because much of the financial, technical, and administrative support has been concentrated in capital cities and with larger administrative bodies, local governments are finding it relatively more difficult to meet the challenges they face.[49]

The situation is further complicated by the limited capacities and capabilities of civil society organizations and an inability to mobilize societal resources more broadly. While NGOs have become a recognized presence in the region, many groups are encountering internal challenges as seasoned NGO leaders step down from their current positions and international sources of funding for the ongoing operations of these groups evaporate.[50] The issue of limited capacity is compounded by historical patterns of reliance on government organs to make decisions and emerging policies within states that restrict participation. These resource and policy trends make it difficult for environmental organizations to be fully integrated into many decision-making processes.[51]

Capacity, Sustainability, and EU Accession

In the period leading up to and immediately following the fall of the communist regimes, environmental issues were at the center of national political agendas. Although domestic environmental issues have shifted to the margins of political concern and debate, the new member states have demonstrated a commitment to adopting the *acquis* and to achieving high environmental standards. While some regional observers suggested that the accession of CEE countries would result in environmental quality being shifted to the lowest common denominator, CEE officials have moved rapidly to enhance national level legal, administrative, and human resource capacities associated with environmental policy and have readily adopted EU directives and regulations. At the same time, the environmental aspects of accession remained important throughout the years-long negotiations process, with EU officials in Brussels growing more serious over time, not less, about harmonization of CEE environmental policies with EU policies.

Traditional capacity building programs have transferred funds, technology, and expertise to support CEE environmental policy and institutions. These programs have targeted state and nonstate actors at different levels. However, to promote long-term sustainability that addresses and responds to ongoing financial and institutional challenges, these programs must be augmented with more process-oriented capacity development initiatives. Capacity development for the environment (CDE) is not and cannot be a one-way, West-to-East process. For countries to meet their environmental goals, both donor agencies and assistance recipients must strive to ensure that CDE initiatives change over time and that they include and empower a wide range of actors.[52] Programs such as PHARE have been transformed over the last decade, moving from funding mostly national-level capacity building programs toward funding more initiatives designed to enhance local and regional public sector capacities. Similarly, other EU donor agencies have assessed the needs of CEE states and societies and adjusted their funding priorities over time. However, to have sufficient capacity to promote and realize sustainable development, CDE activities must continue to anticipate and respond to new challenges that emerge. CEE authorities and societies also face the challenge of raising additional revenues for investment in national and local environmental enforcement personnel, inspection, and institutions. With this in mind, future donor agency CDE might include efforts to "mainstream" and "integrate" sustainability concerns across various donor program areas and work to develop staffing and organizational factors necessary to implement both EU and domestic conceptions of sustainable development.

The wholesale application of EU environmental policy will not "save" the environment in the CEE region or result in the creation of sustainable societies. There are many additional aspects of environmental policy and institutions as well as areas of mass behavior that need to be changed to achieve these lofty goals. While some changes may take place at the EU and global levels, others will require national- and local-level attention across CEE states and societies. The unique characteristics of candidate countries and the variation in their approaches to environmental protection and to enlargement overall suggests that a state-centered approach could provide a basis for uniform implementation of the *acquis* that still promotes national distinctiveness. The nearly single-minded pursuit of EU membership that drove CEE states means that many domestic issues have been subsumed in the EU accession process. With accession agreements finalized and membership attained, such issues are likely to receive more attention. As such, the creation of an environmentally sound and sustainable Europe may require that CEE-EU communication become less unidirectional and more multilateral.

Notes

1. JoAnn Carmin and Stacy D. VanDeveer, *EU Enlargement and the Environment: Institutional Change and Environmental Policy in Central and Eastern Europe* (London: Routledge, 2005); Matthew Auer, ed., *Restoring Cursed Earth: Appraising Environmental Policy Reforms in Eastern Europe and Russia* (Lanham, MD: Rowman and Littlefield, 2004).
2. Petr Pavlinek and John Pickles, *Environmental Transitions: Transformation and Ecological Defence in Central and Eastern Europe* (London: Routledge, 2000); Petr Pavlinek and John Pickles, "Environmental Pasts/Environmental Futures in Post-Socialist Europe," *Environmental Politics* 13, no. 1 (2004): 237–65.
3. This chapter benefits from the presentations and discussions among participants in the workshop on "EU Enlargement and Environmental Quality in Central and Eastern Europe & Beyond" held at the Woodrow Wilson International Center for Scholars in Washington, D. C. in March 2002. We are grateful to the Center for Environmental Solutions at Duke University for providing resources that aided in the preparation of this chapter, and to the Center for Austrian Studies at the University of Minnesota for supporting our participation in its September 2002 workshop.
4. Pavlinek and Pickles, *Environmental Transitions;* Andreas Beckmann, JoAnn Carmin, and Barbara Hicks, "Catalysts for Sustainability: NGOs and Regional Development Initiatives in the Czech Republic," in *International Experiences on Sustainability,* ed. Walter Leal Filho (Bern: Peter Lang Scientific Publishing, 2002), 159–77.
5. These changes are detailed in "Environmental Performance Reviews" of most CEE countries that are organized under the auspices of the OECD and UNECE (cf. OECD, "Environmental Performance Reviews [1st Cycle]: Conclusions and Recommendations: 32 Countries [1993–2000]," OECD Working Paper on Environmental Performance [Paris: OECD, 2000]; United Nations Economic Commission for Europe (UNECE), *Second Environmental Performance Review for Estonia* [Geneva: United Nations, 2001]). A complete list of these published reviews is available at <www.unece.org>. See also "Regular Reports on Progress Toward Accession," especially chapter 22 of each report, at http://europa.eu.int/comm/enlargement/report2001/#Regular%20Reports
6. Christine Kruger and Alexander Carius, *Environmental Policy and Law in Romania: Towards EU Accession* (Berlin: ECOLOGIC, 2001); United Nations Economic Commission for Europe (UNECE), *Environmental Performance Review for Romania* (Geneva: United Nations, 2001).
7. Pavlinek and Pickles, *Environmental Transitions.*
8. Tamar L. Gutner, *Banking on the Environment: Multilateral Development Banks and Their Environmental Performance in Central and Eastern Europe* (Cambridge, MA: The MIT Press, 2002); HELCOM Project Team on Hazardous Substances, *The Implementation of the 1988 Ministerial Declaration on the Protection of the Marine Environment of the Baltic Sea Area with regard to Hazardous Substances,* May 2001; OECD, *Environmental Data Compendium 1997* (Paris: OECD, 1997).
9. Stacy D. VanDeveer and Geoffrey D. Dabelko, *Protecting Regional Seas: Developing Capacity and Fostering Environmental Cooperation in Europe* (Washington, D.C.: Woodrow Wilson Center, 2000). For a detailed discussion of the genesis of capacity building and capacity development, see Merilee S. Grindle, *Getting Good Government: Capacity Building in the Public Sector of Developing Countries* (Cambridge, MA: Harvard University Press, 1997); OECD, *Developing Environmental Capacity: A Framework for Donor Involvement* (Paris: OECD, 1995); Ambuj Sagar, "Capacity Development for the Environment: A View from the South, A View from the North," *Annual Review of Energy and the Environment* 25 (2000): 377–439; Miranda A. Schreurs and Elizabeth Economy, *The Internationalization of Environmental Protection* (Cambridge: Cambridge University Press, 1997).
10. Grindle, *Getting Good Government,* 5.

11. Sagar, "Capacity Development."
12. Susan Baker and Petr Jehlicka, eds., *Dilemmas of Transition: The Environment, Democracy and Economic Reform in East Central Europe* (London: Frank Cass, 1998).
13. Janine R. Wedel, *Collision and Collusion: The Strange Case of Western Aid to Eastern Europe 1989–1998* (New York: St. Martin's Press, 1998).
14. Ans Kolk and Ewout van der Weij, "Financing Environmental Policy in East Central Europe," in *Dilemmas of Transition: The Environment, Democracy and Economic Reform in East Central Europe,* ed. Susan Baker and Petr Jehlicka (London: Frank Cass, 1998), 53–68.
15. Ibid.
16. Ibid.
17. Baker and Jehlicka, *Dilemmas of Transition.*
18. Tom Garvey, "EU Enlargement: Is It Sustainable?" in *EU Enlargement and Environmental Quality: Central and Eastern Europe & Beyond,* ed. S. Crisen and J. Carmin (Washington, D.C.: Woodrow Wilson International Center for Scholars, 2002), 53–62; United States Environmental Protection Agency (USEPA), "Dark Past, Bright Future: Environmental Cooperation in Central and Eastern Europe and the New Independent States" (EPA 160-K-98-002) (Washington, D.C.: USEPA, 2002).
19. See, for example, Michael Toman, *Pollution Abatement Strategies in Central and Eastern Europe* (Washington, D.C.: Resources for the Future, 1994).
20. USEPA, "Dark Past, Bright Future."
21. Toman, *Pollution Abatement Strategies;* Mikael Sandberg, *Green Post-Communism: Environmental Aid, Polish Innovation and Evolutionary Political Economics* (London: Routledge, 1999); Ger Klaassen and Mark Smith, *Financing in Environmental Change in Central and Eastern Europe: An Assessment of International Support* (Laxenburg, Austria: International Institute for Applied Systems Analysis, 1995).
22. Toman, *Pollution Abatement Strategies;* Klaassen and Smith, *Financing in Environmental Change;* Sandberg, *Green Post-Communism.*
23. JoAnn Carmin and Barbara Hicks, "International Triggering Events, Transnational Networks, and the Development of the Czech and Polish Environmental Movements," *Mobilization* 7, no. 3 (2002): 305–24.
24. Barbara Jancar-Webster, "Environmental Movement and Social Change in the Transition Countries," in *Dilemmas of Transition: The Environment, Democracy and Economic Reform in East Central Europe,* ed. Susan Baker and Petr Jehlicka (London: Frank Cass, 1998), 69–92.
25. Ibid.
26. Carmin and Hicks, "International Triggering Events."
27. Henrik Selin and Stacy VanDeveer, "Hazardous Substances and the Helsinki and Barcelona Conventions: Origins, Results and Future Challenges" (presented at the Policy Forum Management of Toxic Substances in the Marine Environment: Analysis of the Mediterranean and the Baltic, Javea, Spain, October 2002); Petr Jehlicka and Andrew Tickle, "Environmental Implications of Eastern Enlargement: The End of EU Progressive Environmental Policy?" *Environmental Politics* 13, no. 1 (2002): 79–95; Baker and Jehlicka, *Dilemmas of Transition.*
28. Accession negotiations between the EU and Bulgaria and Romania have many "open" chapters.
29. Jehlicka and Tickle, "Environmental Implications of Eastern Enlargement."
30. Europa, "Chapter 22—The Environment" (2002). Available online at http://europa.eu.int/comm/enlargement/negotiations/chapters/chap22/index.htm
31. Europa, "Accession Negotiations: State of Play" (2002). Available online at http://europa.eu.int/comm/enlargement/negotiations/pdf/stateofplay_july2002.pdf
32. Kolk and van der Weij, "Financing Environmental Policy."

33. Barbara Hicks, "Setting Agendas and Shaping Activism: EU Influence on Central European Environmental Movements," *Environmental Politics* 13, no. 1 (2004): 216–33.
34. Hicks, "Setting Agendas and Shaping Activism."
35. Carmin and VanDeveer, *EU Enlargement and the Environment.*
36. Gutner, *Banking on the Environment.*
37. Beckmann, Carmin, and Hicks, "Catalysts for Sustainability."
38. John M. Kramer, "EU Enlargement and the Environment: Six Challenges," *Environmental Politics* 13, no. 1 (2004): 290–311 (293).
39. Garvey, "EU Enlargement: Is It Sustainable?"
40. Kramer, "EU Enlargement and the Environment."
41. Cf. OECD, "Environmental Performance Reviews" (2000); Baker and Jehlicka, *Dilemmas of Transition.*
42. Liliana B. Andonova, *Transnational Politics of the Environment: The European Union and Environmental Policy in Central and Eastern Europe* (Cambridge, MA: MIT Press, 2004).
43. Ibid.
44. Stacy D. VanDeveer, "Normative Force: The State, Transnational Norms and International Environmental Regimes" (Ph.D. dissertation, University of Maryland, College Park, MD., 1997).
45. Andonova, *Transnational Politics of the Environment.*
46. Katerine Holzinger and Peter Knoepfel, *Environmental Policy in a European Union of Variable Geometry? The Challenge of the Next Enlargement* (Basel: Helbing and Lichtenhahn, 2000); Sabina Crisen and JoAnn Carmin, eds., *EU Enlargement and Environmental Quality: Central and Eastern Europe & Beyond* (Washington, D.C.: Woodrow Wilson International Center for Scholars, 2002); Selin and VanDeveer, "Hazardous Substances and the Helsinki and Barcelona Conventions."
47. Jehlicka and Tickle, "Environmental Implications of Eastern Enlargement."
48. Ingmar von Homeyer, "Differential Effects of Enlargement on EU Environmental Governance," *Environmental Politics* 13, no. 1 (2004): 52–76.
49. Stacy D. VanDeveer, "Europeanizing Central Europe: Capacity, Surprises, Lessons and Challenges," in *EU Enlargement and Environmental Quality: Central and Eastern Europe & Beyond*, ed. S. Crisen and J. Carmin (Washington, D.C.: Woodrow Wilson International Center for Scholars, 2002), 114–22.
50. Beckmann, Carmin, and Hicks, "Catalysts for Sustainability."
51. Andreas Beckmann and Henrik Dissing, "EU Enlargement and Sustainable Rural Development in Central and Eastern Europe," *Environmental Politics* 13, no 1 (2004): 135–52; Lars K. Hallstrom, "Eurocratising Enlargement? EU Elites and NGO Participation in European Environmental Policy," *Environmental Politics* 13, no. 1 (2004): 175–93.
52. Sagar, "Capacity Development"; Stacy D. VanDeveer and Ambuj Sagar, "Capacity Building for the Environment: North and South," in *Furthering Consensus: Meeting the Challenges of Sustainable Development Beyond 2002,* ed. E. Corell, A. Churie Kallhauge, and G. Sjöstedt (London: Greenleaf, 2006).

Chapter 4

SUSTAINABILITY OF CLUSTERS AND REGIONS AT AUSTRIA'S ACCESSION EDGE

Edward M. Bergman

Introduction

The terms "cluster" and "sustainability" are two of the most provocatively ambiguous in their respective literatures. It may therefore be a risky enterprise to link them in a paper, particularly one that draws attention to development issues already loaded with normative overtones.

However, it seems this is not the first attempt. In proposing a sustainable energy industry cluster for Mesa Del Sol, Serchuk and Singh (1999) ask and then answer "What is sustainability?" After referring to the 1987 Brundtland report of the World Commission on Environment (where sustainability was defined as development by which societies today meet their needs without compromising the ability of future generations also to meet their own needs), they conclude:

> Too often, people interpret this definition exclusively in biophysical terms, and thus concentrate on conservation of physical resources, the protection of human health, the maintenance of stable ecosystems, and so on. We prefer a broader interpretation. Achieving sustainability entails balancing the environment, the economy and social equity. To endure on a sustained basis, a society needs fair access to good jobs and economic resources, and it must factor into its economic accounting the consumption and destruction of environmental resources—especially those that are irreplaceable. *An industry cluster aims to incubate the development of innovative—and therefore robustly competitive—firms* [italics added].[1]

For others, "sustainable development of regions" or "of nations" are equally contested terms: Holmén (2001) considers sustainable development de facto to be an inherently contradictory concept, since "development" itself implies structural change of certain "sustaining" territorial elements, therefore placing initial sustenance at risk.[2]

In expanding the concept beyond biophysical to economic components, innovation and competition receive prominent mention as elements of cluster sustainability. Similar sentiments are reflected in comments by Greene (2001): "The sustainability of the ICT [Information and Communication Technologies] cluster will derive from constant innovation, which in turn must be based on leading-edge research and research training."[3] De Vol (2001) proposes it is worthwhile to consider "what makes some clusters stick while others fall apart? The factors that allowed them to form may not be as important in sustaining them. Especially when many believe that randomness and historical accidents are integral components of how a high-tech cluster starts."[4]

Sustainability Assumed?

Unlike the case of cluster definitions, which are debated at length and with considerable energy,[5] one gets the impression from most published literature sources that cluster sustainability hasn't been considered much of an issue, perhaps because sustainability is an obvious side benefit simply of *having* a successful cluster. The two terms join at the hip in many promotional materials, implying that clients should expect sustainability from clusters, and that attaining the latter virtually ensures the former. Much of the language issued by proud cluster officials during the U.S. business expansion therefore appears celebratory in nature rather than cautiously prudent or forward-looking. Because of these assumptions, the question of sustainability is seldom raised seriously.

To the degree sustainability does receive explicit consideration, ideas seem to revolve more around ensuring "retainability" of *specific cluster incumbents* (firms, institutions, officials) than about sustaining the underlying premises that originally permitted incumbent success. I will return to these underlying premises in later passages.

What I see as a misperception can in part be traced to the predominance of successful clusters among the case studies in the literature. These offer a misleading impression that one need only emulate the set of "best practices," institutions, etc. characteristic of these clusters to achieve success and with it a sustainable future. Researchers tend to overlook cases of stillborn clusters that never succeeded at all (by definition, inherently difficult to identify) or those that succeeded for a time but then failed (easier to find, but just as easily avoided). Tichy (1998) reminds us of many failed clusters available for study, including former precision work clusters (the machine-tool cluster of Baden-Württemberg and Swiss watch cluster), natural resource clusters (coal and minerals, iron, petroleum), and mass production (Detroit auto clusters).[6] Each was in its day a world-class success story, and each suffered massive failure. A few later managed to recapture markedly different or refocused cluster activities.

An article in a recent issue of *Economic Development Quarterly* chronicles the experiences of thirty small U.S. cities that went through similar economic failures and have since recovered to varying degrees.[7] Numerous first-hand accounts taken from different observers attest that local officials were least able to grasp the risks to sustainability during the very period of greatest economic success:

> Everyone was fat and sassy ... everyone was employed ... they frowned upon lower paying jobs[...] The fact that there's always been so many jobs—people could come out of high school and make top wages.... Nobody wanted to change that.
>
> Because of the integration between politics and influential individuals nothing could really happen of a progressive or constructive kind in the city. Any time any one group, like the chamber of commerce, wanted to do something, another group, like the county commission would get their people together and oppose it.
>
> It wasn't until the mid-70s that there was a perception that we had a problem ... big shift in elected officials ... People got very proactive community-wide ... From that point on, it [economic revitalization] has been the number one issue for the entire community.[8]

As we shall see later, the best time to consider questions of cluster sustainability is during their period of greatest success. However, it is at precisely these moments that human nature and local institutions seem least capable of attending to issues more comfortably delayed or relegated to one's successors.

Cluster and Regional Prosperity: Some Common Factors

As has been repeatedly mentioned, the basic ideas behind industrial districts originated at the end of the nineteenth century with Alfred Marshall and Friedrich List, who assessed the significance of such developments for the industrial success of England and Germany, respectively. Subsequent theorists of the early twentieth century such as Weber considered industrial districts to be special cases of economic spillovers—called "localization economies"—that could arise in many regions where firms within a specific industry segment concentrated and prospered.

In both countries, attention was focused on how localities within national economies gain developmental advantages by workforce development and pooling, the reduction of costs associated with agglomerated inputs, adoption of *implicit cluster knowledge*, and the transmission of *tacit craft technologies* developed by and among proximate firms and industries. These were logical extensions of craft-based traditions that preceded industrial revolutions and were carefully nurtured for use within the industrial district.

Many craft-based industrial districts thrive today, mainly in Third Italy[9] firms that continue to rely on the design and craft singularities that ensure market niches among the world's growing middle-class consumers. To the degree clusters resemble these industrial districts, sustainability hinges heavily on successfully following or shaping market tastes and ensuring the exceptionally high design standards and craftsmanship characteristic of niched luxury and discretionary goods. It is in these kinds of districts and using these kinds of skills that proximity, interfirm trust, and cooperation count most. Trust-based cooperation among nominally competing agents is a hallmark of cluster thinking and policy, at least in these kinds of clusters and new industrial districts.

However, harmonious interaction or trust among other types of clusters and firms is seen by Maskell as strictly optional: "The *only* requirement is many firms with similar bodies of knowledge be placed in circumstances where they can monitor each other constantly, closely and almost without effort or costs."[10] Rosenfeld (2001) documents this type of minimal—even hostile—interaction among furniture manufacturers in a northern Mississippi cluster that "steal each others' designs and attempt to produce them at lower cost ... recruit each other's employees for small increases in wages,"[11] or among a cluster of houseboat builders in south-central Kentucky where "competition is so fierce that some companies are not willing even to be present in the same room with others ... [which] stems primarily from owners of new startups who ... left established companies ... and took with them either employees, customer contracts or both."[12] Maskell's underappreciated point is that even in exceptionally competitive—virtually combative—circumstances, beneficial knowledge and innovation spillovers can arise without need for agreements, cooperation or much trust among cluster firms. This is yet another way to understand how both cooperation and competition can coexist among firms in differing cluster formations, and it also suggests why some interdependent firms strongly resist formal organization or governance even as they enjoy the benefits of voluntary clustering.

Traditional localization economies and pecuniary externalities found in industrial districts described above work well for firms and industries facing stable technologies and traditional best practices. However, global markets and widely spread production or distribution facilities alter our understanding of how other clusters are exposed to technological possibilities that affect globally originating risks and opportunities of many kinds. According to Maskell, "It was, however, only towards the turn of the [twentieth] century that the advantages stemming from knowledge creation and knowledge spillovers occupied centre stage in the conversation on the cluster...."[13] In the mid 1990s, these factors led OECD to consider the innovative role industrial clusters play in major work programs concerning national innovation systems,[14] which culminated in two published volumes, the first of which "accomplished the valuable and vital task of broadly informing OECD readers about value-chain concepts and the potential of clusters that function as reduced-form national innovation systems."[15]

The reasons for broadening original industrial district ideas to clusters of national and international dimensions are by now the stuff of everyday news: emergent communication technologies, explosion of Internet effects, lowered transportation costs (mass air travel, inexpensive air freight, transport deregulation), international intellectual property rights protections, emergence of NAFTA, EU, and WTO trading regimes. These and all other increasingly familiar technologies and institutions permit rapid and effective diffusion of formal, novel, and commercially applicable technologies through various corporate channels, global supplier-chains, licensing and patent agreements, and business-to-business technical services.

Localities that remain insulated from these high-energy channels of knowledge transfer and commercial contact risk losing touch with the next generation of competitive technologies. At the same time, firms in localities or clusters that are not subject to global competition will remain unaware of changing customer demands that firms in more competitive surroundings are fully prepared to meet. These are among the reasons another OECD-led initiative argues that: "Clusters have attracted widespread attention as potential motors of economic growth and social innovation, allowing SMEs [Small and Medium size Enterprises] to compete on a global scale. This is why the topic of clusters is of great importance in the context of transition economies where a productivity gap persists and where SME growth remains feeble at best."[16]

Simultaneously, new competitive technologies also support effective global deployment of facilities owned by multinational firms seeking strategic location advantage. It is perhaps here that the broad scope of Michael Porter's work during the last decade assumes its greatest importance: he argues that corporate strategic planning should be relatively less concerned about operational efficiencies of individual units and rather more focused on capturing strategic competitive advantage. This is best accomplished by locating corporate units in the most hospitable clusters and regions of the world. The difference between operational efficiency and strategic advantage mirrors a similar difference between what are called static and dynamic externalities, where the latter, according to Henderson et al., deal "with the role of prior information accumulations in the local area on current productivity and hence employment."[17]

In his widely respected business publications on how to evaluate regions that meet strategic business needs, Porter coopted (with surprisingly scant acknowledgement) much of the existing cluster and facility location literature written originally from the perspective of regional strategists. He further adapted it to corporate settings through (perhaps incessant) use of the famous "diamond" of corporate success factors. It is often the case that indigenous startups or spinoffs of independent firms are initially responsible for putting hospitable regions on the maps made available to the corporate strategists Porter advises. One highly desired spillover sought when making facility location decisions is entrepreneurial energies, which frequently accelerate following corporate investment in promising regions.

In my view, the greatest value Porter adds to cluster concepts is his *linking* of globally driven competition (and supporting technologies, often new and disruptive) and the strategic corporate search for existing industrial districts that offer highly differentiated localization economies of unique benefit to subsets of international firms and functions. It is this combination that most properly bears the label "Porterian clusters." Moreover, his linking has also helped expand the concept of clusters well beyond locally bounded industrial districts. It is now common to speak of national and supranational or mega and macro clusters, all of which reflect the highly elastic, nested relationships characteristic of global trade and communication technologies.[18] Sustainability of clusters formed by Porterian (and Marshallian) forces depends heavily upon access to global sources of formal knowledge and explicit technologies—often embodied in capital goods or knowledge workers' human capital—and their timely diffusion to local clusters and districts.

Cluster Life Cycles

The risks to cluster sustainability consist of life-cycle rigidities, which have been studied to determine their effects on development paths or the life-cycle of regional economies[19] and clusters.[20]

Two interesting life-cycle perspectives are worth exploring further. Tichy argues that cluster sustainability is a matter most properly considered over a fairly long developmental wave, during which a cluster first forms, then grows rapidly and reaches maturity or perhaps terminal petrification. He models this process closely on a product cycle theory analog, where formative clusters are at their most technologically dynamic phase, acquiring and upgrading skilled workers, deploying and refining new techniques or commercializing new products, and seeking new markets and customers. Developmental histories of clusters resemble those of their host regions in several but not all respects, since one can be relatively strong while the other is relatively weak.

In contrast to the abundance of studies on firms, surprisingly little has been written about a cluster's formative phase,[21] perhaps because clusters are seldom recognizable sufficiently early to outside observers, although an identification of deliberate actions leading to formative stages has been advanced elsewhere.[22] Therefore, we are not entirely sure about processes of early spillovers, the emergence of trust and reciprocity, and the formation of lasting networks, infrastructure, and so on, a process that mirrors Maskell and Malmberg's view of how regional capacities emerge: "The localised capabilities are all moulded by historical processes."[23]

Tichy considers it possible to create clusters in the presence of existing network strengths (three are identified as key: labor, input-output, and technology networks). Policy-created clusters are said to be more likely to succeed when they are *network-based* (with cooperative production of range of services, similar stocks of knowledge, different supply chains or customer branches) rather than *star-clustered* (with a dominant firm and its linear/dedicated supplier chain, similar to growth-poles), although the latter are faster and their startup costs much lower.

The possibility of deliberate cluster creation contrasts with the views of many who see the emergence of "organic" or "evolutionary" clusters as serendipitous accidents of historical accumulation wherein regional resource endowments, assorted "Porterian" corporate strategists, entrepreneurs, and innovators, or simply felicitous coincidence, could feature prominently. Organic clusters contrast rather sharply with consciously deliberate regional efforts to design and launch the necessary networks. There are smatterings of evidence to support both views, with European experience more interventionist and American experience driven more by markets, chance, and economic history.

Maskell and Malmberg argue that isolating mechanisms are at work in innovative regions, sustaining them by protecting them from external regional competitors.[24] First, *asset mass efficiency* is the idea that historically agglomerated research and development (R&D) and related innovation assets are not easily or readily duplicated in competing regions. Second, *time compression diseconomies* are the costly but necessary lags a competing region must overcome while trying to master and replicate the capacities of a superior region, which can busily continue to build upon its strengths through increasing returns processes. Last, an externally inscrutable *interconnectedness of asset stocks* implies that simply replicating each asset stock produces no sense of how they are deployed effectively, which is another way of saying that accumulated assets develop DNA-like usage patterns not visible or apparent to outsiders or even those who daily draw upon this embedded DNA.

Returning to Tichy's view of the development wave or life-cycle concept, the growth phase of a cluster

> appears to be the best of all worlds to participants. It is the phase, nevertheless, which may generate the first deviations [that] cause later troubles. Success is easy in this phase, so that little pressure exists to search for further development of the cluster's strengths, for other applications of its knowledge, etc. It is tempting to concentrate on the best-selling product(s) and to produce it (them) in ever-increasing quantity, utilising economies of scale. As a consequence economic policy must stand against concentration, try to avoid overspecialization, and protect the region's information density.[25]

The structural risks any region faces during periods of rapid growth are in such moments too bothersome or confidence-challenging to deal with, yet it is during precisely such times that decisions that could place cluster sustainability at risk are inadvertently made.

To avoid structural risks during the growth and maturation stages, Tichy argues that regional officials should stop promoting R&D dedicated to a cluster's main specialities, encourage other related uses of a cluster's dominant skills and products, promote regional spinoffs and related research capacity, encourage multiple producers while discouraging single producers of given cluster products, and build foundations that encourage expansion of new clusters to broaden a region's overall cluster portfolio.

Maskell and Malmberg's regional version of processes associated with maturation considers *asset erosion* first, which in their view takes place as "hitherto important institutions in a region are no longer reproduced at the same pace or to the same degree."[26] A second factor, *substitution*, is a special form of technological erosion that arises when "new technology rapidly devalues former investments in, for instance, skills, education and infrastructure, thus undermining the region's capabilities." Others term these innovations "disruptive technologies," which first find acceptance only in marginal market niches but then, through design/production refinements and shifts in customer demand, eventually displace the originally "sustaining technology." Last, *regional lock-in* can develop when initially important institutions and practices—often social and cultural in origin—focus on self-preservation or aggrandizement and become a sclerotic risk to[27]—rather than the lifeblood of—regional progress.

The mature phase, which is critical in Tichy's view, arises when a cluster's potential to react to market demand and develop new offers is nearly exhausted. This means all three underlying networks have lost the capacity to adapt, collapsing instead upon a highly specialized set of redundant skills, obsolescent process and design technologies, and mass-produced output lines attractive to ever fewer customers and supply chains. The Swiss pharmaceutical cluster is said to have fallen into this state, now producing only very traditional products, even though parent Swiss pharmaceutical *firms* have managed to keep up with the demanding medical product cycle by conducting their most critical operations—Porter-style—in various U.S. and other advanced-economy regions. The remaining shards of network competence in a mature cluster that has become unfavorably positioned might still be redirected to regional emerging clusters where their contributions could be beneficially deployed, but otherwise the older cluster is headed for terminal "petrification," the endgame Wilbur Thompson once termed an "industrial hospice."

A final petrification phase arrives when all remaining cluster activity is conducted by a single failing entity, perhaps an endangered branch or nationalized firm, which operates with the certain knowledge it cannot long survive, even for routine political purposes of providing subsidized or regulated employment for its remaining workers. So much effort and resource expenditure may have been required to retain maturing and petrifying cluster remnants and resuscitate exhausted cluster networks that the region will have placed its final bets on the wrong cluster and paid very high opportunity costs in the bargain. In such a case, the possibility of sustainability has long since passed and the host region could be placed at considerable risk, absent outside beneficial forces.

Sustainability Lessons: Innovate Constantly, Learn by Competing

The literature makes clear just how dependent clusters and regions are on processes that expose them to competition in demanding global markets and to sources of innovative practices. This may not be news to many, but still one might ask: Are firms' capabilities to "innovate constantly, learn by competing" enhanced more in clusters or in regions?

Relying mainly on the typical cluster literature discussed earlier, one cannot avoid the indelible impression that clusters *per se* are automatically innovative and therefore sustainable, although the literature on regions is considerably less optimistic, with regions known to be in dire situations far outnumbering permanently prosperous regions that enjoy innovation-based success. One is therefore quite logically led to a naive hypothesis: sustainability is more likely in the average cluster than the average region.

The remaining sections of this chapter attempt to explore this hypothesis in the context of Austria and its CEE neighbours, starting with a description of how the relative strength of Austria's clusters and regions is determined. The rest of the chapter then presents a stylized set of innovation and competition indicators for clusters vs. regions of varying strength ratings, ending with a short summary of the major findings and possible implications. The findings are of considerable conceptual interest, but are also important to accession countries and regions now considering alternative development concepts that compete directly for scarce material resources and time, which if misallocated in periods of "path dependency" could exact high opportunity costs.

Cluster and Region Strength

A fair comparison of innovation and competition in regions vs. clusters should account for varying degrees of cluster or regional strength. Ideally, we should also seek systematically recorded evidence of such practices and perspectives, rather than piece together the usual patchwork of often conflicting, contradictory, or incommensurable case-study results.

Accordingly, this section will report on a survey of Austrian firms conducted in 2001 to assess their innovative capacities and competitive instincts; all of the firms are located in areas that adjoin one of Austria's four EU accession-country borders.[28] A key feature of the survey includes a self-assessment by each firm of the cluster *and* region within which the firm operates. Distinctions between cluster and region make little sense in case studies of monocluster regions—such as the Third Italy industrial districts—and the fact that regions and clusters could vary considerably in their practices is routinely ignored in other case studies of both clusters and regions.

Moreover, when cluster and regional concepts are collapsed for analytic convenience, evidence concerning essentially different processes that may account for variations in cluster vs. regional development becomes confused or overlooked. To avoid further propagation of these and similar problems, each firm reported in this study was asked to rate the strength of its region on a five-point scale: poor-to-excellent regional innovation and investment environment; clusters were also rated on a five-point strength scale from emergent to developed cluster firms and institutions.[29] A firm could rate both cluster and region as high- or low-strength, meaning there is no practical difference between their strength levels (i.e., they are highly rank correlated), or it could rank them independently (no rank correlation), implying wholly distinct strengths for the two environments experienced by a given firm. The Spearman rho correlation of .203 reveals some similarity of rankings, but the similarity is not convincingly robust.[30]

A cautious reader might note that firms are inclined to report their clusters and regions to be of similar strength somewhat more frequently than they report differing strengths. If strength similarities are overreported by Austrian firms, then the weakly correlated strengths of clusters and regions become weaker still, which calls into serious question the convenient assumption, held by many scholars and analysts, that clusters and regions function as close substitutes or proxies.

Sources and Uses of Innovation

It is widely agreed that firms are best able to survive and prosper in competitive economies such as the EU and NAFTA when they draw upon various sources of proven technological and commercial knowledge that permit them to innovate successfully. The innovation systems literature points in particular to various stages of innovation where firms might logically deploy knowledge and identifies typical sources of such knowledge. Early stages include (1) the generation of potentially valuable ideas and (2) the subsequent development of such ideas prior to their commercialization.

The sources of potentially valuable knowledge within an innovation system are usually associated with basic knowledge institutions, such as universities, plus intermediary organizations that transfer knowledge to members, or with the accumulated commercial knowledge embedded within the local-to-international communities of firms, particularly other firms in the same industry, customer firms, and supplier firms.

Based upon the innovative inputs that our respondent firms report deriving from each of these sources, we are able to detect some useful distinctions between strong regions and clusters at the two product innovation stages.[31] The sources of innovative inputs are grouped in rows of Table 4.1 according to the following categories: *supportive institutions* (business incubators and industry or cluster associations), *same sector firms* (associated with horizontal clusters or localization economies), *regional firms* (value-chain suppliers or customers), *venture capitalists*, and *universities*. The columns of this table indicate whether the innovative services benefited firms at the *idea generation* or *idea development* stages, where generation of innovative ideas implies a head start. What we are looking for is evidence that certain sources or stages of innovation become more important for stronger clusters or stronger regions. This will be indicated by rank order correlations (Φ) between cluster (C) or regional (R) strengths and a particular innovative source.

Table 4.1 Sources of Innovative Inputs

Innovation Sources/Stages	Supportive Institutions*	Same Sector Firms	Regional Firms*	Venture Capitalists	Universities
Idea Generation	—	RΦ= .00 CΦ= .28	RΦ= .24/.28 CΦ= .18/.20	RΦ= .22 CΦ= .00	RΦ= .38 CΦ= .18
Idea Development	RΦ= .00/.00 CΦ= .32/.52	RΦ= .00 CΦ= .35	RΦ= .21 CΦ= .36	RΦ= .18 CΦ= .22	RΦ= .34 CΦ= .34

* indicates more than one source of innovation and associated correlation in the category

One notes immediately that as cluster strength grows, firms show generally higher tendencies to rely upon all sources for idea development; regional strength associations with nearly every idea development source are much weaker or absent, the exception being universities, which are equally important for both regions and clusters. Conversely, idea generation gains importance more for rising regional than cluster strengths, the exception being same-sector firms, which are often considered *de facto* evidence of a cluster.

Where supportive institutions (incubators and cluster organizations) and same-sector firms provided greater innovation assistance at both stages to stronger clusters, venture capitalists and universities provided more assistance to the strongest regions. Regional firms (value-chain partners) provide stronger regions with idea generation services and stronger clusters with idea development services, which captures wholly the preceding paragraph's principal observation.

Taken together, stronger clusters depend decisively upon institutional, same-sector firms and regional firms for idea development support, while stronger regions depend more upon regional firms, universities, and venture capitalists for their idea generation assistance. Perhaps strong cluster firms are simply able better to absorb and exploit organizational inputs for idea development services. Alternatively, strong cluster firms have become so dependent upon organized and shared innovation sources that they are less able or inclined to tap innovative generation potentials in their regional community of firms, venture capitalists, and universities.

One is tempted to interpret these patterns as partially refuting our working hypothesis and appearing more consistent with the alternative view that strong regions have greater sustainability prospects than strong clusters, which are heavily reliant on closely associated institutional or cooperative arrangements rather than on the market or knowledge-based sources that characterize strong regions. The consequences for sustainability hinge on the answer to this question: Are stronger cluster firms likelier to engage in sustainable activities when innovation is increasingly drawn from supportive institutions and same-sector firms, or are firms in stronger regions more likely to engage in sustainable practices? The evidence of competition, markets, and cluster advantages will be examined to help sort these matters out a bit.

Cluster Advantage vs. Competitive Markets

Part of the answer we seek can be found in the degrees to which strong clusters and strong regions are able to capitalize upon a series of advantages that often characterize industry clusters. These are examined in Table 4.2, whose rows represent a series of potential advantages and whose columns contain the rank correlation (Φ) for firms located in strong regions (R) and clusters (C).

Strong regional and clustered firms alike acknowledge these advantages, but those in clusters clearly identify more strongly with more of the potential advantages: the weakest cluster correlation approximates the strongest region correlation. This imbalance is the direct result of rapidly diminishing advantages to regions at their highest strength levels, because lower strength levels account for all the observed correlation. In contrast, every cluster advantage rises in importance among firms across the full low to high cluster strength pattern, some quite markedly. The attenuated regional strength pattern contrasts clearly with the steadily increasing advantages of the cluster strength pattern, which consequently enjoys measurably stronger correlations.

This implies that firms attached to clusters of highest-rated strength continue to draw more heavily upon potential cluster advantages, no matter the strength of their region. The strong connections offer indirect confirmation of the conceptual consistency between cluster strength rankings and known attributes of clusters.

The earlier reviews of sustainability remarked that any cluster or region could risk losing its competitive edge over time due to many factors that will not be rehearsed again here. However, the many factors can be boiled down to two: (1) innovative practices (e.g., idea generation and development) that permit renewal of capacities and options, as evaluated previously, and (2) competitive pressures on firms to differentiate themselves, which usually stimulates the pursuit of innovation, thereby reducing the risk of complacency and sclerotic closure, a discussion of which now follows.

To get at the possibilities of competitive differentiation, we focus attention on how successfully respondent firms are able to position themselves within various market regimes and where they obtain knowledge that permits such positioning. In Table 4.3, five groups of responses are used to demonstrate contrasts among clusters and regions in their orientation toward competition and differentiation.

Table 4.2 Strong Cluster and Strong Region Advantages

Cluster Advantages	Strong Regions	Strong Clusters
Member relationships support R&D efforts	RΦ= .26	CΦ= .60
Associations/institutions promote success	RΦ= .34	CΦ= .53
Regional firms prefer to work together	RΦ= .25	CΦ= .31
Knowledge is gained from cluster firm contacts	RΦ= .24	CΦ= .46
Firms participate in cluster-wide programs	RΦ= .00	CΦ= .33
Firms/institutions are open to entry of new firms	RΦ= .00	CΦ= .49

Table 4.3 Orientation of Clusters and Regions toward Competition and Differentiation

Positioning Tendencies of Firms	Strong Regions	Strong Clusters
International Market Positioning		
—Demanding international customers	RΦ= .17	CΦ= .00
—ISO certification	RΦ= .19	CΦ= .00
—Proportion of production exported	RΦ= .23	CΦ= .00
Regional Market Positioning		
—Demanding regional customers	RΦ= .18	CΦ= .00
—Regional customer feedback	RΦ= .28	CΦ= .00
—Regional supplier quality	RΦ= .29	CΦ= .00
Specialized Differentiation		
—Specialized supplier assistance	RΦ= .19	CΦ= .19
—Degree of product difference with strongest competitor	RΦ= .17	CΦ= .22
Buyer trends gained as cluster member	RΦ= .21	CΦ= .41
Competition costs outweigh benefits	RΦ= .00	CΦ= .24

Here again, we are looking for evidence of how firms position themselves as the region or cluster of which they are part varies from weak to strong. This is important, since we want to know if firms are either better able or more willing to position themselves competitively as their regions and clusters become more fully developed and stronger.

Starting first with *international market positioning,* firms may position themselves in response to sophisticated and demanding customers who raise the competitive bar through their willingness to buy, thereby pressuring firms to innovate more. Alternatively, the need to obtain ISO certification (International Organization for Standardization) of products and processes to compete effectively in global markets calls forth innovative responses. Finally, a higher percentage of total production exported to international markets (e.g., Latin America) demonstrates an ability to meet international competition in the globalizing economy. As we can see, firms find these positioning elements important as the strength of their host regions rises, but the same firms show indifference to such positioning factors over the full range of observed cluster strengths.

Next, we look at *regional market positioning,* where firms are locally subject to sophisticated customer demands, receive regional customer feedback regularly, and are able to deploy the specialized inputs of high-quality regional suppliers. These are often considered some of the most significant factors behind the success of both regions and clusters, but only those firms in regions of increasing strength show any relationship. In other words, firms hosted by stronger regions enjoy greater degrees of these features, but the same firms in stronger clusters report no such relationship.

Another positioning technique is the firms' *specialized differentiation* of their products from their competitors', thereby ensuring some temporary protection and pricing advantage as well. In the effort to differentiate products or services, firms often collaborate with key suppliers to develop new offers. Firms tend to seek identical degrees of assistance from specialized suppliers to differentiate product lines as their host regions *or* clusters become stronger. Together with other innovative practices discussed in earlier sections, firms assemble their product portfolios, which differ from their strongest competitors' to various degrees, as reported here. Firms appear to have slightly more differentiated product lines in strong clusters vs. strong regions.

The ability to differentiate product lines and to anticipate new market opportunities often depends upon collective information on the *buyer trends* that are developing. Membership in a cluster provides access to such collective information, which may help firms anticipate new markets earlier and more effectively. Although firms in strong regions and strong markets appear to benefit from information concerning new buyer trends, the degree of association is nearly twice as high for firms in strong clusters than strong markets, which helps confirm the relative advantage of collective product market information.

Finally, local competition is often said to stimulate firms to become more efficient, innovative, and successful in both local and international markets. The question is, do firms in local markets see the advantages of local competition, thereby encouraging such competitive behavior, or do they see the *costs of competition* outweighing the benefits? This is a key factor in the development of clusters and regions, since competition appears to be an important component of sustainability. The evidence here is quite striking: firms in regions of all strengths hold no systematic view, i.e., responses were essentially random, but firms in stronger clusters systematically find that the costs of competition outweigh the benefits. Perhaps competition is seen as harmful to good relations within "cooperative" clusters, or perhaps firms become steadily more averse to competition as cluster strength and services grow; whatever the cause, efforts to avoid competition in strong clusters may lead to serious sustainability problems.

A consistent and revealing portrait of differences between strong region and strong cluster firms gradually becomes clearer: strong region firms are far more likely to rely on market contacts and regional economic strengths to remain competitive. On the other hand, strong cluster firms are much likelier to rely on organizational services and membership benefits to become competitive. Strong region firms would therefore have a natural competitive advantage, since they rely most heavily upon market processes.

Conclusions

This chapter started with an inquiry into the sustainability of clusters and regions, where it was argued that sustainability depends upon innovative sources of renewal and upon competitive pressures that stimulate or provoke firms to innovate valuable new products or processes.

Drawing upon recently collected evidence from Austrian manufacturing firms that self-assessed the strength of both clusters and regions to which they belonged, some important differences emerged. First, cluster and region strength is somewhat similar, though not identical, for firms. Cluster features and advantages rose steadily for firms as their clusters' strength increased, but these features and advantages peaked early and did not rise further for firms as their regions' strength increased. A cluster "advantage plateau" was clearly reached early in the development phase of regions.

Second, innovative sources differ for cluster vs. regional strengths, where strong cluster firms draw their most important innovative inputs from organizations and strong region firms rely far more on other regional firms. Third, competitive markets and production processes, combined with strong customer- and supplier-based product development assistance, clearly distinguish firms in strong regions. Strong cluster firms were highly dependent on cluster and industry organizations to remain competitive and expressed less competitive concern as cluster strength rose.

One is, therefore, led to the tentative conclusion that strong regions are probably more sustainable than strong clusters. Regions may in fact have been evaluated as strong by a respondent firm if they promote sustaining practices, i.e., they are places where innovation and competition are embedded in the very economic fabric. On the other hand, cluster firms' heavy dependence on formal organizations could make them more vulnerable to the sclerosis that creeps into unchallenged bureaucracies and self-perpetuating structures. Equally suspect is the growing aversion of strong cluster firms to local competition, which reveals a development mechanism prone to serious sustainability risks. (These findings are summarized in Appendix II.)

These comments do not imply that cluster development should be avoided, but they do suggest that clusters may be of more limited usefulness than presently thought.[32] For example, the known advantages firms enjoy as their clusters gain strength seem to flatten out early in regional strength rankings. This seems to imply that clusters might play key *supplemental* roles in weak regions, providing important services, effects of scale, and synergy found lacking in the region. However, the value of cluster supplementation to firms may diminish as regions develop, advance, and agglomerate "naturally" so that adherence to a cluster may inhibit firms from becoming valuable members of a complex and sustainable region.

This view appears to be broadly consistent with recent research in the U.S., which found that urbanization externalities (complex "Jacobs" regional effects) promoted sustained growth, unlike localization externalities ("MAR" [Marshal-Arrow-Romer] industrial concentrations/clusters).[33] If so, one might be led to consider how to promote simultaneously the development of clusters *and* regions, refocusing attention gradually away from clusters as they become fully integrated into sustainable regional economies.

The experience of Austrian firms may prove instructive to neighboring accession countries and regions as the many development options available in the literature or presented for consideration by EU, OECD, and other officials come under serious review.

Appendix I

Representative Definitions

—An *industrial district* is "a socio-territorial entity, characterized by the active presence of both a community of people and a population of firms in one naturally and historically bounded area."[34]

—A *cluster* is a "geographically-bounded concentration of interdependent businesses with active channels for business transactions, dialogue and communications, and that collectively shares common opportunities and threats."[35]

—"A *cluster* is a geographically proximate group of interconnected companies and associated institutions in a particular field, linked by commonalities and complementarities. The geographic scope of a cluster can range from a single city or state to a country or even a network of neighboring countries."[36]

—"*Clusters* can be characterized as networks of production of strongly interdependent firms linked to each other in a value-adding production chain ... [with particular reference to the] concept of *economic clusters* as a reduced scale model of innovation system approach."[37]

Appendix II

Sustainability Findings and Conclusions

1.
Strong *Clusters* → member advantages gain steadily
Strong *Regions* → member advantages plateau early

2.
Strong *Clusters* → organizations support innovation
Strong *Regions* → regional firms support innovation

3.
Strong *Clusters* → membership aids competitive position
Strong *Regions* → market exposure drives competition

4.
Strong *Clusters* → costs seen as greater than benefits of competition/rivalry
Strong *Regions* → indifferent to local competition/ rivalry

5.
Strong *Regions* more sustainable than Strong *Clusters*
Strong *Clusters* potentially important in Weak *Regions*

Notes

1. Adam Serchuk and Virinder Singh, "Sustainable Energy Industry Cluster for Mesa Del Sol," Renewable Energy Policy Project, 1999. http://www.repp.org/repp_pubs/articles/mesaDSol/index.htm
2. Hans Holmén, "The Unsustainability of Development," *International Journal of Economic Development* 3, no. 1 (2001): 1–26.
3. Roy Greene, "Ireland's winning industrial formula protects economy against effects of US downturn," University of Galway Press Release, 15 May 2001.
4. Ross de Vol, "Talents are attracted by the quality of Place: Connected Clusters Stick," *BrainHeart Magazine* (March 2001): 8.
5. See Appendix I for representative definitions and sources.
6. G. Tichy, "Clusters: Less Dispensable and More Risky than Ever," in *Clusters and Regional Specialization,* ed. M. Steiner (London: Pion, 1998), 211–25.
7. H. J. Mayer and M. R. Greenberg, "Coming back from Economic Despair: Case Studies of Small- and Medium-Size American Cities," *Economic Development Quarterly* 15, no. 3 (2001): 203–16.
8. Mayer and Greenberg, "Coming back from Economic Despair," 208–9, 213.
9. "Third Italy" denotes northeast central Italy, a region of scattered industrial districts that prospered economically because of the vitality of its small, clustered firms, in stark contrast with southern Italy and the major industrial centers of Italy.
10. P. Maskell, "Towards a learning-based theory of the cluster" (paper presented at the World Conference on Economic Geography, 4–7 December 2000, Singapore), 14.
11. S. Rosenfeld, "Advancing the Understanding of Clusters and Their Opportunities for Less Favored Regions, Less Advantaged Populations, and Small and Mid-Sized Enterprises" (Carrboro, NC: Regional Technology Strategies, Inc., 2001), 8.
12. Rosenfeld, "Advancing the Understanding of Clusters and their Opportunities," 10.
13. Maskell, "Towards a learning-based theory of the cluster," 8.
14. "National innovation systems" or NIS is a concept or framework that sets forth the institutional and organizational features of how a nation or a region systematically generates, uses, and replicates its knowledge and innovation base.
15. P. den Hertog, E. M. Bergman, and D. Charles, eds., *Innovative Clusters: Drivers of National Innovation Systems* (Paris: OECD Proceedings, 2001), 7, referring to T. J. A. Roelandt and P. den Hertog, eds., *Boosting Innovation: The Cluster Approach* (Paris: OECD Proceedings, 1999).
16. Local Economic and Employment Development (LEED), *Clusters in Transition Economies—Progress Report,* DT/LEED/DC(2002)8, p. 2. The OECD report continues: "It needs to be said at the outset that to date, no clusters—defined as spontaneously occurring agglomerations of vertically and/or horizontally integrated firms operating in related lines of business—have yet been identified in [Slovenia, Slovakia, Czech Republic, Poland, and Hungary]," p. 3. The report is available at http://www.oecd.org/dataoecd/49/22/2089148.pdf
17. V. Henderson, A. Kuncoro, and M. Turner, "Industrial Development in Cities," *Journal of Political Economy* 103, no. 5 (1995): 1067–90.
18. P. den Hertog, E. M. Bergman, and D. Charles, "In Pursuit of Innovative Clusters" (paper presented at International Conference on Measuring and Evaluating Industrial R&D and Innovation in the Knowledge Economy, 23–24 August 2001, Taipei, R.O.C.), 6–9.
19. Bergman, E. M., and Harvey Goldstein, "Urban Innovation and Technological Advance in the Research Triangle Region," in *Sustainability of Urban Systems: A Cross-National Evolutionary Analysis of Urban Innovation,* ed. P. Nijkamp (London: Gower, 1990); P. Maskell and A. Malmberg, "Localized Learning and Industrial Competitiveness," *Cambridge Journal of Economics* 23 (1999): 167–85.
20. Tichy, "Clusters: Less Dispensable and More Risky than Ever."

21. "Formative phase" refers to the actual emergence of firms and supporting institutions that enjoy cluster benefits à la Marshall, not the issuance of nomenclature, labels, and websites favored in recent years by many a newly established institutional presence that wishes to stake formal claims of existence.
22. M. Peneder, "Dynamics of Initial Cluster Formation: The Case of Multimedia and Cultural Content," in *Innovative Clusters: Drivers of National Innovation Systems,* ed. P. den Hertog, E. M. Bergman, and D. Charles (Paris: OECD Proceedings, 2001), 303–12.
23. Maskell and Malmberg, "Localized Learning and Industrial Competitiveness," 173. Continuing: "The present built environment and infrastructure can often be traced back through at least a century, while the natural resources typically are from pre-historic origins. It should, furthermore, be noted that the resources available in the region comprise the region's own resources and the ones available through import from other parts of the world."
24. Maskell and Malmberg, "Localized Learning and Industrial Competitiveness," 176–78.
25. Tichy, "Clusters: Less Dispensable and More Risky than Ever," 233.
26. Maskell and Malmberg, "Localized Learning and Industrial Competitiveness," 178–9.
27. M. Olson, *The Rise and Decline of Nations: Economic Growth, Stagflation and Social Rigidities* (New Haven: Yale University Press, 1982).
28. The survey includes seventy-eight usable responses from a sample of industrial firms stratified by four value chains (vehicle producers, electronics producers, chemical and pharmaceutical producers, and metalworking and machine producers) and four major production regions (Vienna, Graz-Villach, Klagenfurt, and Linz-Wels) in 2001. The same sampling frame was earlier used to assess the contingent valuations of transportation and logistics services by a similar sample of firms in the same value chains and regions; G. Maier, E. M. Bergman, and P. Lehner, "Conjoint Analysis of Transport Options in Austrian Regions and Industrial Clusters," in *Freight Transport Demand and Stated Preference Experiments,* ed. R. Danielis (Milan: Franco Angeli, 2002). The questionnaire was developed and refined by the author, in conjunction with ongoing OECD cluster studies, drawing upon other instruments that have been used to examine clusters and innovation. Parts of it have also been used in more focused studies of U.S. and Slovak regions.
29. The Austrian firms were asked to respond to the following questions:

Figure 4.1 Survey Form for Austrian Firms (2001)

Considering all the significant factors, including government, industry and social factors, how good a location is your region as a place *to innovate* in your business?		
Very poor location in which to innovate or invest further	1 2 3 4 5	Excellent location in which to innovate or invest further

Your industrial cluster…		
Is still emerging, with a narrow range of firms and institutions	1 2 3 4 5	Is well developed with a broad range of firms and institutions

30. The two-tailed test of this correlation remains insignificant at the 5% level, while the one-tailed test passes the 5% but not the 1% confidence level. Readers should use their own judgment in assessing these exploratory results, some of which are of marginal significance.
31. In all comparisons, we are principally interested in detecting whether respondent firms are more likely to engage in or support a variety of sustainable practices as the strength of their association with (1) regions or (2) clusters rises. In other words, do stronger regional or cluster contexts improve chances of sustainability equally? Unless otherwise stated, significance is measured with one-tailed tests, where the majority of tests are significant at 0.05 or better ("~" indicates significance at 0.10 or better).

32. These results could also indicate that Austrian firms ranked their *efforts* to stimulate clusters, rather than the actual functional effectiveness or performance level of clusters of industries. If so, the difference may also reflect similar differences concerning EU-favored cluster launching efforts vs. U.S.-favored *ex post facto* support of market-based agglomeration clusters, which are relatively understudied but quite interesting issues.
33. E. L. Glaeser, H. D. Kallal, J. A. Sheinkman, and A. Scheifler, "Growth in Cities," *Journal of Political Economy* 100 (1992): 1126–52.
34. Becattini, G., "The Marshallian ID as a socio-economic notion," in *Industrial Districts and Inter-firm Co-operation in Italy,* ed. F. Pyke, G. Becattini, and W. Sengenberger (Geneva: International Institute for Labour Studies, 1991), 37–51.
35. S. A. Rosenfeld, "Bringing Business Clusters into the Mainstream of Economic Development," *European Planning Studies* 5, no. 1 (1997): 3–23.
36. M. E. Porter, "Competitive Advantage, Agglomeration Economies and Regional Policy," *International Regional Science Review* 19, no. 1 (1996): 85–94.
37. Roelandt and den Hertog, *Boosting Innovation,* 9.

Part Two

THE ECONOMICS OF SUSTAINABLE DEVELOPMENT

Chapter 5

GREENHOUSE GASES EMISSIONS TRADING IN THE CZECH REPUBLIC

Jiřina Jilková and Tomáš Chmelík

Introduction

The Czech Republic has to implement a domestic trading system as part the EU emission trading system. This instrument is viewed as a pillar of a cost-effective framework not only to reduce greenhouse gas (GHG) emissions at home, but to help to improve energy efficiency and further reduce air pollution.

This paper presents basic ideas and discussions about a trading system with greenhouse gases in the Czech Republic, incorporated in a complex instrument framework for climate change and energy efficiency issues. The analysis of issues is seen in the scope of a regional context (other Central and Eastern European countries) and the global context.

Energy Developments and Policy

The Czech Republic belongs to the Central and Eastern European countries (CEE). Usually, this term is identified only with the former socialist bloc; nevertheless Vienna is farther to the east than Prague. All CEE countries with experience of communist rule are now EU accession countries, undergoing intensive restructuring and fundamental political changes.[1] The development of energy policy and climate change issues in the Czech Republic can serve as a typical case to illustrate the problems of the situation in transition economies. Before the political changes in 1989, the CEE countries were centrally planned economies.

Absolute priority was given to production targets, with no account taken of environmental impacts. Economic development was highly focused on the expansion industries, which were heavily polluting.

The character of the centrally planned economies, with no private ownership and no functioning market, did not create economic incentives for efficient use of natural resources, including energy. While political representatives publicly advocated conservation of nature, energy conservation, and energy efficiency, no funds were provided for energy efficiency investments. The technologies used were obsolete, and pollution abatement measures were hardly undertaken at all. The adverse environmental effects on the region were compounded by the consumption of very low-quality domestic coal, in the absence of oil and gas resources in the region. The main energy source, lignite, has the lowest carbon content of all coals.

There continues to be a strong need to improve energy efficiency in the CEE region, for the following reasons:

- To be competitive in the EU and world markets,
- To improve the foreign trade balance by lowering the high imports of energy,
- To comply with the requirements for entry into the EU,
- To fulfill international commitments and protocols concerned with air pollution control,
- To reduce the adverse environmental effects of energy use.

The driving forces of the energy markets in CEE countries are energy price reforms and the removal of energy subsidies.

Figure 5.1 compares the total primary energy supply (TPES), expressed as a share of GDP, using exchange rates, of the Czech Republic and the Slovak Republic with various regions and neighboring countries. The graph clearly shows that transition countries still have significantly higher energy intensity related to domestic product, even if a positive trend can be observed.

Figure 5.1 Total Primary Energy Supply (TPES/GDP, toe per thousand 1995 USD, exch. rate)

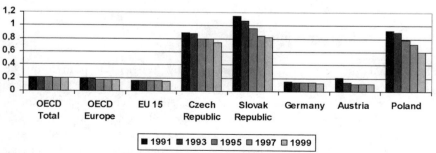

Source: International Energy Agency (IEA), 2001.

Figure 5.2 Total Primary Energy Supply (TPES/population, toe per capita)

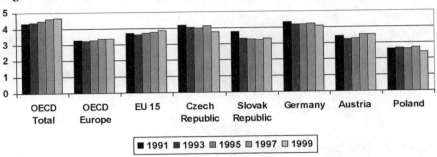

Source: IEA, 2001.

Figure 5.2 shows that primary energy supply per capita is very similar to developed countries' (often even lower). The problem that CEE transition economies are facing is low productivity, so that a similar population produces much less product. This results in similar figures of energy intensity per capita, but much worse energy intensity per unit of GDP. In the Czech Republic, the energy mix has become less dependent on solid fuels, but coal, the main source of power generation in the country, still remains at about 49 percent of TPES. The economy is still characterized by relatively high primary energy consumption (167 giga joules [GJ] per capita) and a high level of GHG emissions.

Climate Change Issues

Since the early 1990s, negative environmental impacts from the energy sector and air pollution have gradually been reduced in postsocialist CEE countries, thanks to the implementation of new environmental laws with performance standards, large investments in environmental protection and new technologies, and energy intensity improvements. However, emissions of air pollutants per capita (and per GDP) in economies in transition still remain higher than in developed countries (see Table 5.1). The NOx (nitrogen oxides) emissions are more influenced by the growth in transport, which is similar in all countries.

The driving force of postsocialist CEE countries is the transition from a planned to a market economy. Most of them are likely to have an emissions surplus in the first commitment period of the Kyoto Protocol (2008–2012), possibly with the exceptions of Slovenia and Croatia. The projected surplus varies greatly, depending on future development.

Both energy and environmental policies are directly linked to the EU accession process. These policies have played a role in the reduction of GHG emissions, even if they were implemented for a variety of reasons other than GHG abatement.

Table 5.1 Air Pollution in Selected Countries (1997)

Indicator	Czech Republic	Austria	Poland	Slovakia	EU15	OECD EURO
SO_x kg per capita	45	7.1	21	62	25.5	29.8
NO_x kg per capita	42	25	26	24	30.3	28.2
CO_2 toe per capita	12.6	7.4	8.7	8.4	8.9	8.2

Source: MOE CR, *Statistical Environmental Yearbook of the Czech Republic* (Prague: The Ministry of the Environment of the Czech Republic, 2001).

The higher level of energy intensity in transition economies suggests greater opportunities for emissions reductions. Some countries see advantages in early action in emissions trading, because it will promote industry involvement and provide a learning effect. In most CEE countries, joint implementation is more developed than emissions trading.

In the Czech Republic, GHG emissions have been reduced in the last ten years by approximately 26 percent of the emissions balance in 1990, mainly because of the transformation of the Czech economy. However, CO_2 and SO_x emissions per capita are still higher than the EU average. Table 5.2 and Figure 5.3 indicate the development of CO_2 emissions from the relevant sectors. Emissions from energy production make up the main part of total emissions, with the most rapid reduction occurring between 1990 and 1999 (MOE CR and Czech Hydrometeorological Institute [CHMI] 2001).

For the purpose of emissions projections and analysis of mitigation options, emissions scenarios were elaborated in 2000.[2] One set of scenarios describes "business as usual" (BAU), while the other should reflect the impact of mitigation policies and measures. Three different BAU scenarios have been prepared to project country emissions for a time span reaching only to the end of the first commitment period. These scenarios depend mainly on the intensity of economic growth (ranked as low, medium, or high).

Figure 5.4 charts a basic comparison of the real inventory trend with the three levels of BAU scenarios by 2010. It can be seen that there is a very realistic possibility of reducing GHG emissions in the Czech Republic by at least 14 to 34 percent relative to the 1990 level. This leads, at a minimum, to an estimated potential of about 60 to 240 Mt of CO_2 for possible international emissions trading in the five-year period 2008–2012. It is obvious that introducing domestic emissions trading into the national climate change strategy could even raise this rough estimate.

These trends indicate that the CR is expected to be in compliance with its Kyoto target under its current set of policies. It is likely to be a seller on the international emissions trading market, although the precise amount is still very

Table 5.2 Emissions and Sinks of CO_2 from Relevant Sectors, 1990–1999 (Mt CO_2)

	1990	1991	1992	1993	1994	1995	1996	1997	1998	1999
Combustion	160.1	148.8	135.6	130.7	123.6	124.6	130.3	132.1	125.1	118.6
in energy production					61.4	66.6	57.9	59.4	59.1	54.4
in industry					33.4	30.1	43.9	43.3	35.4	34.2
in transport	8.0	6.9	8.1	8.3	8.3	8.9	10.4	11.8	11.0	12.6
of which road transport			7.3	7.4	7.4	8.0	9.3	10.9	10.3	11.3
all other fuel combustion	34.9	28.7	22.7	21.6	20.6	19.0	18.2	17.6	19.6	17.5
of which mobile fuel combustion sources			1.5	1.1	1.3	1.0	0.6	1.1	1.1	1.3
Industrial processes	5.4	4.3	4.6	4.2	4.1	4.2	2.5	2.5	2.7	2.4
Other sources							0.5	0.5	0.7	0.7
Total emissions	165.5	153.1	140.2	134.9	127.7	128.8	133.3	135.1	128.5	121.6
Sinks in forestry	−2.3	−5.0	−6.0	−5.6	−3.9	−5.5	−4.5	−4.6	−3.8	−3.4
Emissions minus sinks	163.2	148.1	134.2	129.2	123.8	123.4	128.8	130.4	124.7	118.2

Source: MOE CR and CHMI, *The Czech Republic's Third National Communication* (Prague: The Ministry of the Environment of the Czech Republic and Czech Hydrometeorological Institute, 2001).

Figure 5.3 CO_2 Emissions from Relevant Sectors 1990–1999 *(Mt CO_2)*

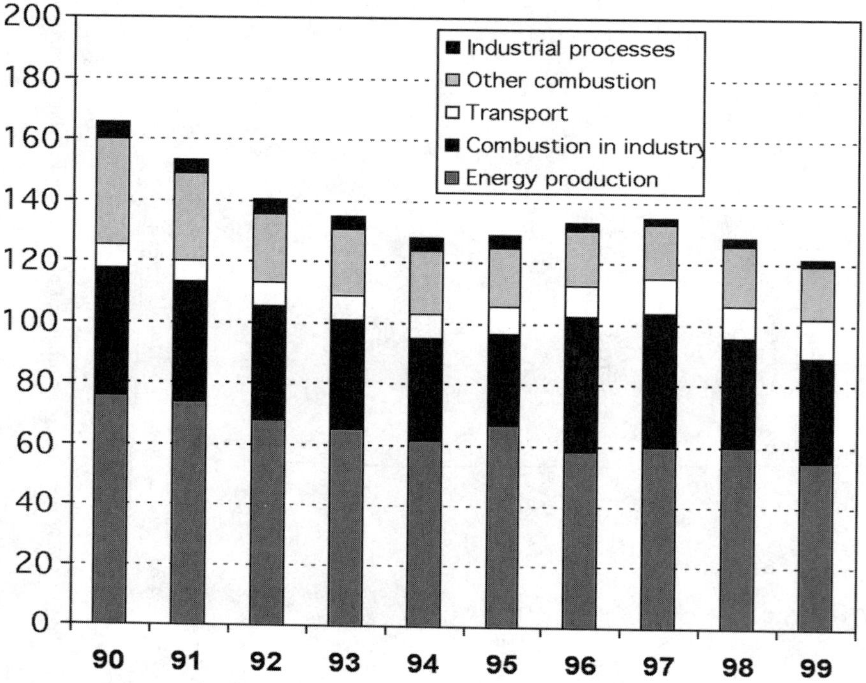

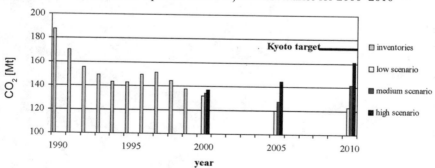

Source: MOE CR and CHMI, *The Czech Republic's Third National Communication* (Prague: The Ministry of the Environment of the Czech Republic and Czech Hydrometeorological Institute, 2001).

Figure 5.4 Inventories and Updated BAU Projection Scenarios for 2000–2010*

*GHG emission in 2000 (preliminary data): 141.8 Mt.

Source: CHMI and SRCI, PHARE Project, 2000.

uncertain. However, the CR still has an interest in further reducing GHG emissions, for a variety of reasons: supply of GHG permits to the international market as well as secondary benefits, such as reduced air pollution, increased energy efficiency, and lower energy dependency.

Environmental policy implemented in the 1990s has contributed to large reductions in air pollutants through emissions standards and fiscal instruments. Emissions standards were mainly met by the installation of scrubbers in the energy sector, but also led to a reduction in coal use in favor of natural gas, thereby reducing GHG emissions. Environmental policy is now beginning to recognize climate change as a priority issue, in particular due to the accession process. The Czech Republic's State Environmental Policy now includes the objective of reducing GHG emissions by 20 percent in 2005 compared to the 1990 level. Revisions of the Clean Air Act incorporated climate change considerations.

The national climate change strategy, approved in 1999, formally accepted the national greenhouse gas emissions reduction target set in Annex B of the Kyoto Protocol, a reduction of 8 percent of GHG emissions in the period 2008–2012 compared to 1990. The strategy establishes the respective responsibilities of government bodies. It also provides a general orientation for the implementation of policies and measures: the highest priority is placed on policies and measures aimed at emission reductions at the national level. Additionally, the emissions reductions could be transferred internationally through the Kyoto mechanisms.

The existing policy reflects a broad spectrum of measures related to energy saving and increased penetration of renewable energies. There is, however, no strategic approach to GHG emission reductions. Measures often are introduced separately and aimed at different primary targets, with GHG reductions occurring as a side effect. It is expected that a new strategy should include a clear goal for emissions reduction and instruments directly emphasizing GHG emission reductions, as well as better coordination and linkages between instruments.

New Policy Options

The driving forces for the design of new policies were the EU accession process, the EU proposal for the directive on emissions trading and renewable energy, the energy intensity and air pollution situation, and the Kyoto Protocol.[3]

Climate change and energy policy generally represent a type of problem where experts or representatives of different departments of the Ministry of Environment (MOE) or even different ministries or agencies have to be involved. The example of the Czech Republic could point out some problems that occur in other CEE countries in connection with the implementation of flexible instruments (or with climate change issues generally).[4] Analysis of the current situation shows the need to devise a system that will combine the existing framework of measures with new instruments attuned to development and challenges in their national and international context.

The system proposed consists of the following main instruments:

- Emissions trading as the pillar of the system, accompanied by and linked with:
- Air pollution charges and ecological tax reform,
- Subsidies (e.g. support of renewables and energy savings),
- Project-based measures (Joint Implementation, Prototype Carbon Fund),
- IPPC Directive,
- New CO_2 provisions in the Air Protection Law,
- The Waste Management Law for CO_2, CH_4, N_2O regulation.

Emissions trading and joint implementation under the Kyoto Protocol are defined as flexible instruments. In the medium-term perspective, the goals of the program are:

- GHG emissions reduced by 20 percent from the 1990 level (high emissions per capita and GDP),
- Possible link with the EU trading scheme,
- Proof of concept and demonstrations of the feasibility and merit of the domestic system as a low-cost way to reduce GHG emissions,
- Getting practical experience in emissions trading, so that Czech entities can access the international trading systems (learning by doing),
- building effective institutions and standardized procedures.

In the long-term perspective the goals of the program are:

- Sustainable emissions reductions beyond the first commitment period,
- Identification of least-cost opportunities for emissions reductions,
- Revealing real marginal abatement costs for Czech entities, which can then get better information in advance about their opportunities on the international market.

The limited effect of existing policies and measures and the advantages and benefits of potential emissions trading predestine this emissions trading scheme to set new and effective incentives to improve the impact of other instruments and to reduce further GHG emissions in the Czech Republic.

Joint Implementation

Most CEE (EIT) countries prefer Joint Implementation (JI) at the moment. There are different reasons, not least among them the potential benefits of being a host country (lower abatement costs, administrative structures in place) and the sense that there is no need to reduce GHG emissions at the moment because of the surplus situation. Meanwhile, potential investors bring significant pressure to act as a host country and implement JI projects.

The Czech Republic stayed for a relatively long time in the AIJ (Activities Implemented Jointly) phase, during which five AIJ projects were realized. In January 2002, the start of a JI phase was officially announced together with basic information about the Ministry of the Environment's approach to the issue:

- For JI projects approximately 2 Mt CO_2 eq. annually has been allocated, i.e., 10 Mt CO_2 eq. in all in the period 2008–2012.
- Only investment projects or groups of projects that comply with the valid legislation of the Czech Republic may be proposed as JI projects. Simultaneously, the project must not lead to the transfer of pollution between the individual components of the environment (air, water, soil). An already realized project cannot be approved as a JI project.
- The following areas are priority areas for JI projects:
  - utilizing renewable energy sources,
  - energy savings in heating buildings (insulation, regulation) in the public sector,
  - energy savings in heating apartment buildings (insulation, regulation),
  - utilizing waste industrial heat in existing installations,
  - construction of collection systems for landfill gases in old landfills and use of the energy therefrom,
  - ecological public transport.
- Other installations leading to a substantial decrease in GHG emissions shall also be eligible.

Guidelines and Administrative Structures

The submitted project must include the following basic information, according to which the project will be evaluated:

- Transparent calculation of the baseline (reference level) of GHG emissions; for newly built sources of heat and energy, the reference value shall be set hypothetically,
- Annual reduction of GHG emissions,
- Total reduction of GHG emissions during the project lifetime,
- Amount of emission credits required in 2008–2012,
- Price of an emission reduction unit (investment/emission credits),
- Economic effectiveness of the project,
- Other environmental effects.

In addition, the following general criteria shall be assessed when evaluating the project:

- The condition of "additionality," i.e., the decrease in GHG emissions resulting from the given technology that would be impossible to achieve without

implementing the project. Proposals for projects that are aimed at fulfilling conditions laid down by the relevant environmental legislation valid in the Czech Republic shall be excluded from further proceedings;
- Compliance with the priorities of the State Environmental Policy and with the priorities of the State Program of Support for Energy Savings and Use of Renewable Sources of Energy;
- The condition of "the best available technology";
- The potential for dissemination of know-how and new technologies in the Czech Republic;
- Compliance with macroeconomic policies, at both national and regional levels (e.g., growth of employment).

In parallel with these basic requirements, the administrative structure for assessing JI project proposals has been discussed and proposed. Its elements can be described as follows. The project proposal is presented to the Ministry of Environment of the Czech Republic, where it receives an identification number. The proposal is then given to the Czech Energy Agency and State Environmental Fund for evaluation. Both of these institutions already have experience evaluating projects focused on energy savings and utilization of renewable energy sources, and their teams of experts are able to assess all the main environmental and economic criteria. The result of the evaluation will be a common position in the form of a background document for meeting a MOE working group with a proposal for a decision (yes or no, and if yes, under which conditions, etc.).

However, a number of issues and responsibilities in the whole process have yet to be decided. On the one hand, the first experience with methodical guidelines showed that the system substantially improved the availability of information and clearly presented the track that the Czech Republic intends to follow.[5] On the other hand, a number of problems appeared as well, so in sum the experience resulted in a relatively significant redesign of the document, which is currently in the process of preparation. Some of the problems are described in more detail below.

One of the problems that was uncovered during the discussions about the approval procedure for JI projects was the question of responsibility or authorization for official approval of JI projects. The problem has two different levels. First, it must be absolutely clear who is competent to authorize the transfer of any emissions units to another authority. At the internal level, we discovered some uncertainties with regard to the Ministry of Environment being the sole authority empowered to give such approval or authorization. This problem will have to be solved at the level of government. It has been suggested that the authorizing entity would be the Ministry of the Environment (i.e., the minister).

A second problem relates to the actual realization of JI projects. At the beginning, in harmony with the language of the Kyoto Protocol and related documents, it was believed that a JI project would be realized primarily between two parties, both parties being countries (one the host country with a potential for cheaper emissions reductions, and the second the investor country with a demand for such reductions). Either country could authorize a private company, an organization, or any other agency to prepare and realize the JI project, but the final agreement about the transfer of financial means and emissions reductions would be made at the official level. For such purposes, memoranda of understanding covering some general issues would be prepared, followed by more detailed agreements for each project or group of projects. In this case, both governments would have a chance to include their national priorities or requirements for potential JI projects in the memoranda or agreements.

Later, however, we recognized the private sector's interest in investment in JI projects (on a totally private basis). Such companies or institutions would enter a JI project as investors, under the same conditions as any other JI project. The difference would lie in the contractual assessments of such projects. Because there would be no memoranda of understanding (between governments), all the issues would have to be covered within the project agreement itself. Also, responsibilities and guarantees would have to be defined a bit differently than in the case of the government agreements. Generally, the treatment of responsibilities is to one of the most sensitive issues in both cases. The problem is that the specific nature of a JI is defined between parties, so that initially there was the idea that an umbrella agreement between parties (Memorandum of Understanding) is necessary to authorize the whole transaction. But if a private company enters a JI, it can get the ERUs, hold them, and use them later for its own compliance (in case it is allowed to fulfill commitments of individual entities with JI credits or to enter these credits into any trading scheme at the company level) or sell them to any other party (company) later on. It will probably be required to have any transfer of credits authorized by a central authority of the host country (to control the conditions under which credits are transferred, ensure environmental integrity, fulfill domestic emissions targets, etc.). However, the form of this approach to financing will have to be subjected to discussion (simply said, if everything turns out well with the project, there should not be a problem on our side; in other words, whether the investor is a party or a private entity should not make a difference from the host country's point of view). Other, related problems of a more or less formal nature and matters of protocol will have to be resolved as well. The final agreement about the transfer of credits will probably be signed by an authorized person (a minister most probably, on our side) whose counterpart in such a case will probably be a representative of a firm.

Portfolio-type Projects

Soon after the JI phase was started, it became clear that there were not enough projects sufficiently large to be economically efficient in the sense of fixed costs related to project preparation. We discovered that in the Czech Republic the potential of "big enough" projects delivering satisfactory emissions reductions "at one place" is relatively limited; however, there is reasonable potential for small-scale projects (e.g., small municipalities' district heating, "house" measures, etc.). Because these projects address some environmental issues that other measures are not able to address, supporting them is a priority. Generally, due to their size, they suffer from a chronic dearth of financial means for investment, so carbon financing could be one of the ways to promote improvements at this level.

Another problem is that official representatives have very limited experience with pooling financial resources, particularly foreign ones, including lack of experience with administrative procedure, language problems, etc. One of the ways to counter this problem is to prepare a portfolio project that encompasses a number of same or similar smaller individual projects, thus sharing the fixed costs associated with a project's preparation, baseline calculation, verification, etc. The Czech Ministry of the Environment is supportive of this approach; however, we soon discovered that our methodical process for evaluating these projects is not adjusted to such an approach. The problem is one of evaluation, which must cover economic and environmental aspects of the project (every single project is evaluated) in a complex way.

Cofinancing

It was (and partially still is) believed that carbon financing in JI projects could be an important, stimulating effect in their realization. This is because projects delivering additional emissions reductions are not economically efficient enough to have return rates of commercial size (if they were, the carbon financing would not be needed). One of the advantages of JI and participation by the central authority in this issue is that through the establishment of a set of priorities the JI can be oriented toward areas where additional financing is needed. Another benefit is that targets of other policies can be met as well (regional development, support of new technologies, support of unemployment, etc.).

The problem is that the typical areas stated for JI projects, such as energy savings, use of renewables, and energy efficiency, have to be to a certain extent supported (subsidized) by a state or other sources (usually public), because they are not competitive on a market basis. This is obvious in the case of measures in the municipal or public sphere (municipal central heating, for example, where the social dimension is one of the key factors). For such purposes, most of the countries have subsidy or support programs that provide soft loans, grants, or other types of financing to make these projects or measures economically feasible.[6]

It is obvious that priority areas for JI projects do cover or collide with some support programs of these agencies. The crucial question is, if the project is financed partially by its implementer and could be eligible for support requested from one of these agencies to cover the rest, where does carbon financing fit? Should it contribute to the financing by the project implementer (and so reduce its own contribution of sources), or should it reduce state support, or should it be divided between both sources? The second version is good for the state as a provider of the support, because it saves financial resources for other projects, but this solution gets no support from the project implementer—the extra costs and work related to preparing the project as a JI do not change the amount of money he has to invest into the project. So, if his financial participation is the same and the "state support only" approach (in comparison to reduced state support plus the JI component) is easier to prepare, he would logically have absolutely no interest in preparing the project as a JI.

On the other hand, if the JI component is fully used to reduce inputs by the implementer while the state support remains the same as without a JI component, the question of additionality arises: would the project be realized anyway if state support remained unchanged? Actually, the additionality issue was discussed from another angle. The idea was to use projects waiting in the "pipeline" for support from the State Environmental Fund as a pool of potential projects for JI. However, the foreign investors later rejected this approach because an independent validator could have problems with the additionality issue—given the "risk" that projects would be realized anyway, no additionality would be required. To deal with this issue we decided to arrange our evaluation and approval procedure so that first the project is evaluated and approved as a JI; then it can (under certain conditions emerging from the project being a JI) enter the regular support programs (for utilizing renewables, for example) of the State Environmental Fund or any other source. The situation is complicated by the fact that the price offered per unit of reductions is relatively low, which makes the carbon financing (price per ton multiplied by reductions achieved) a less important segment of the whole financial side of the project (somewhere around 5–10 percent of total investment costs).

A related and important problem is the fact that such projects have their most acute need for money during the investment phase, which is sometimes difficult to negotiate. Investors, of course, prefer payment upon delivery. In such a case, carbon financing can improve the operation and the later payback period of an already constructed or realized project. Is there additionality at all? Can we say that without carbon financing, even if it enters the financing of the project at a later stage, the project would not occur, especially in a case where financial programs are in place to support the same types of projects and a number of already realized and operating projects have been supported before—all without carbon finance? These issues naturally have a daunting political dimension, but nonetheless represent a pressing problem. All these issues will be subject to intensive discussions in the Czech Republic.

Emissions Trading

In opposition to Joint Implementation, which is already being developed in most CEE countries in transition, emissions trading has long languished under discussion. Only in Slovakia did a specific emissions trading scheme for SO_2, based on quotas already introduced, come into force in January 2003. Intensive preparation of CO_2 emissions trading is also under way here.

As explained above, most postsocialist CEE countries have a surplus of allowable GHG emissions (hot air). The implication is that they are in compliance with the Kyoto Protocol and will stay so even if their economy grows. Emissions trading is an instrument for resolving the problem of scarcity in allowable emissions, so there is no clear reason to introduce emissions trading for the purpose of achieving environmental targets.[7]

The main driving force for the implementation of emissions trading is EU accession. In October 2001, the European Commission published a proposal for an emissions trading scheme within the EU.[8] The implementation of the system proposed by the commission requires each member state and accession country joining the EU to introduce an emissions trading scheme based on a national allocation plan for emissions.[9] The challenges and problems that most accession countries face are different from those in most EU member states. The trading system has to manage problems in terms of the use of scarce resources (one scarce resource is the capacity to absorb GHG). Accession countries (EIT countries) have no scarcity problem; they have free access and a free lunch. They have to introduce the emissions trading scheme in line with the *acquis* implementation and to benefit from the expected and desired learning effect. Early participation of firms could help EITs prepare to enter the international GHG trading market during the first commitment period, gain economic and environmental benefits, and prepare the ground for more substantial emissions reductions in the longer term.

In the Czech Republic, there is currently no emissions trading scheme in place. In 1996, a study on the possibility of using this instrument to reduce sulfur dioxide (SO_2) was carried out.[10] The project covered a single district, Sokolov, located in Northern Bohemia, and focused on issues related to implementing this instrument on such a local basis, including some model simulations that proved an approximately 15 percent reduction in overall costs in reaching a given emissions target in comparison to other instruments. No concrete steps toward practical implementation of the trading system were made.

The design of a cap-and-trade system should reflect the specific circumstances of a transition country like the CR. Since the CR will most probably be in compliance with its Kyoto emissions target under business-as-usual conditions, the goal of such a system, in particular during the precommitment period, would be to gain experience in operating this kind of system.

The Czech Republic can be characterized as a typical representative of an accession and EIT country. The higher level of energy intensity there means that there are greater opportunities for emissions reductions, although there is also a great variance between countries that depends largely on the extent to which the baseline is mainly coal or includes also renewable energy sources, including nuclear power.

The EU system requires mandatory participation of defined sources. As discussed in the previous section, the Czech Republic is expected to have a surplus in the first commitment period. In the period 1990–1999, a reduction of 26.6 percent in GHG emissions was achieved. The estimated reduction by 2010, relative to the 1990 level, is 14–34 percent, which in turn puts the estimated emissions potential in 2008–2012 at 60–240 Mt CO_2. These facts point to different problems and conflicts involved in working out the reduction target and allocation scheme in the Czech Republic.

In the context of the EU proposal for an emissions trading directive, the following points and questions have to be discussed and analyzed:

1. How to set the national target, which will determine the amount of the national emissions surplus.

When actual emissions are less than the commitment, it becomes much more difficult to define volumes of emissions and the time period for trading, because there is no urgent need to take any action. The arguments in support of such a country introducing such an instrument are thus also more difficult, as there is no real target to be achieved. Burdening the national economy with stronger targets than are really necessary might negatively affect competitiveness and the performance of the economy as a whole. Also, political and public support for such measures would be scant.

Yet as we have said, the reason for establishing the trading system is not the requirements of the proposed EU directive alone. A functioning trading system is also needed to develop practical experience in CO_2 trading in preparation for a carbon constrained future. Setting the target emissions goal in the context of initial allocation will be the key point of political discussion.

2. Who should be eligible to sell the surplus (will it be allocated to the state or to companies)?

This question has high political sensitivity. The state would like to do it, but this would erase any stimulative effect for companies. The only means of pleasing and convincing major stakeholders is to hold out the possibility of sharing the benefits of the surplus by dividing its use between the state and the companies.

3. How should the allowances to installations be allocated (what is the allocation scheme)?

This problem is related to the previous question, and to the rules for state aid. The allocation scheme (national allocation plan) will determine the distributive impacts of the whole system. Limiting the eligibility to participate to a small number of firms does not significantly affect the system coverage, defined in terms of GHG emissions covered by the system (as a percentage of total GHG emissions). In the CR, a few sources are responsible for a high proportion of emissions. Energy-related CO_2 emissions from 141 sources (over 50 MW) represent about 47 percent of all GHG emissions in the CR. Increasing the number of sources only increases coverage in terms of GHG emissions by a few percentage points.

However, further studies will need to evaluate whether this leads to loss of market efficiency or to distortions of competition within economic sectors. Sources are defined in terms of emissions and energy output. More analyses are needed to determine the economic sector to which each firm belongs and whether there is a risk of distorting competition among firms belonging to the same sector, if only a subset of these firms are included in the system. This may also raise legal issues, e.g., those related to the EC competition law.

Initial allocation based on a simple rule like grandfathering may have a very different effect from firm to firm, taking potential growth of the firm as one of the most important factors. Some might benefit as net sellers, while others may need to buy emission permits or undertake emissions reduction measures. This seems somewhat inevitable in a trading system; the alternative, designing individual targets for each company, would certainly be a very complex task with uncertain results. However, it might be worthwhile to look more carefully at current emissions projections for individual sectors so as to analyze the likely effects of different cap levels and formulas for the initial allocation. Providing overlarge windfall benefits to some firms might be interpreted as an unfair subsidy. This is particularly true if the system is voluntary and if only those firms that expect to benefit from joining in the system will participate. The final decision on the initial allocation of emissions rights can reflect further analyses and calculations (e.g., using historical data over an average of several years).

Other Policies and Measures

In the context of climate change and energy intensity issues, emissions trading and joint implementation address only large sources and selected sectors. To cover the full spectrum of concerns, these instruments have to be accompanied by a set of instruments focused on problems not covered by flexible instruments. A long-term strategy is needed to provide industry with secure goals, establish a stable entrepreneurial environment allowing strategic decisions, and address the main environmental problems in a coordinated way.

In some CEE countries, these issues are regulated by a climate change strategy. The official document on the State Environmental Policy of the Czech Republic (MOE CR 2000) addresses these issues in a fragmented and uncoordinated way. Many of the stated steps and measures have only a declarative character, without quantitative targets. There is a lack of coordination in policy between the environmental, energy, and agricultural spheres. The real environmental policy is driven both by international agreements and, domestically, by the particular interests of certain groups and sectors, with a focus on administrative and fiscal instruments and their possible revenues.

Conclusions

Analysis of the development in climate change policies at the international and national level shows the necessity of long-term strategic adjustment in environmental policy. The design and implementation of domestic emissions trading systems is under discussion in a number of countries and international institutions. The main driving force behind this is the implementation of the EU emissions trading scheme.

The Czech Republic, a typical representative of transitional CEE countries, is expected to comply with the Kyoto Protocol and to have a surplus of GHG emissions. The directive for an EU-wide emissions trading system has established new perspectives for climate change policy. It creates the potential for significant economic revenue coming from flexible mechanisms that can bring additional financial resources for environmental protection to transitional CEE countries.

However, there is no strategic approach in place to address the issues of climate change. Strategic and economic considerations warrant establishment of the complex framework of instruments, with emissions trading as a pillar, needed to enter the international GHG emissions trading system. The pilot program for the precommitment period should test the concept of emissions trading in the CR's specific national circumstances, discover the real prices and mitigation costs, and help to build and test effective institutions and standardized procedures. Finally, emissions trading and joint implementation as project-based trading should be accompanied by additional measures like abolishing energy subsidies and introducing energy taxes.

Notes

1. In the following text, we speak about the postsocialist CEE countries' economies in transition (EITs).
2. CARLBRO/DHV CR/SRCI CZ. *Estimate of the Economic Costs for the Reduction of Greenhouse Gas Emissions*. PHARE No. CZ 9705-05-05-01-02. Prague: CARLBRO/DHV CR/SRCI CZ, 2000. The letter groups in the "author" slot are the official names of three private companies that prepared the report.
3. The political uncertainty of the Kyoto Protocol is the reason for its limited importance in policy design in CEE countries.
4. A Working Group on Climate Change has been established at the Ministry of Environment for the purposes of discussing climate change issues within the MOE. The WG is chaired by the director of the Department of Environmental Economy and the participants include the Departments of Air Protection, Strategies, Integrated Financing, Global Relations, the State Environmental Fund, the Czech Hydrometeorological Institute, and the Czech Environmental Inspection. Other experts can be invited too, but without the status of a member, they cannot vote. The WG serves as an advisory body to the minister and, among other duties, it also suggests (or refuses) JI project proposals (for administrative structure, see below). The WG meets on no regular basis but rather as necessary.

    However, a number of issues require discussion or approval with other ministries. For such purposes there has been an Interministerial Working Group on Climate Change, where representatives of other ministries, agencies, and the Czech parliament are invited. The group is chaired by the deputy minister of Environment. Due to its "high-level" status, the group can serve only as a platform for discussions or approval of strategic issues or decisions and is able to meet a few times a year only.

At the Ministry of Industry and Trade, a Working Group for the Energy Strategy of the Czech Republic has been newly established.

5. MOE CR, *State Environmental Policy* (Prague: The Ministry of the Environment of the Czech Republic, 2000). The document contains definitions of basic terms and relations, information about the submission and approval procedure, responsible institutions in this process, priority areas for JI, and criteria to be used in evaluation. It also describes in detail all the documents and information that are required for evaluation of the project proposals.
6. In the Czech Republic, there are two agencies that mainly cover this area: the Czech Energy Agency, focused on energy savings, and the State Environmental Fund, focused on renewable energy sources. The types of support are various, depending on the type of measure, the entity asking for support (public/private), etc. Generally speaking, pooling different resources is possible; sometimes it is required that a certain percentage volume of public sources not be exceeded. Part of the financing must covered by the project implementer's own sources; this share is usually smaller in the case of public entities, where in some selected cases the support can even be a 100 percent subsidy.
7. The accession countries are likely to have a surplus in the first commitment period with the exceptions of Croatia, Slovenia, and possibly Lithuania. Croatia and Slovenia are likely to have a slight deficit due to high emissions growth in the early 1990s, and Lithuania could possibly face a deficit if its large nuclear power plant is closed down.
8. European Commission, *Proposal for a Directive establishing a framework for greenhouse gas emissions trading within the European Community,* COM(2001) 581 (Brussels: European Commission, 2001). http://europa.eu.int/eur-lex/en/com/pdf/2001/en_501PC0581.pdf
9. The EU emissions trading scheme was implemented with the directive 2003/87/EC.
10. Jirina Jilková, *Economic Instruments for Large and Medium Sources of Air Pollution—Tradable Permits Pilot Project.* Study prepared for the Harvard Institute for International Development under a cooperative agreement with the U.S. Agency for International Development (Prague: privately printed, 1997).

Chapter 6

ECOLOGICAL REFORM IN THE TAX SYSTEM IN POLAND

Olga Kiuila and Jerzy Śleszyński

This chapter concentrates on ecological tax reform, regarded as an advisable environmental policy option for Poland. Hypothetical effects of the ecological tax reform are assessed using computable general equilibrium modeling. Generally, results confirm both theoretical and practical suggestions from the ecological tax reform experiences in Western European countries. The best results are obtained when the tax burden is shifted from the labor force or household income to environmental pollution. Analyzed scenarios show that results are very sensitive to specific assumptions adopted in the model and that a double-dividend solution is hardly possible. However, long-run effects of the proposed tax reform in Poland would improve welfare and would not slow down economic growth when the appropriate scenario was implemented. Comments on the environmental policy in Poland indicate, indeed, the need for changes in the entire tax system.[1]

Introduction

During the 1990s, several countries (Denmark, Finland, Germany, Italy, Netherlands, Norway, Sweden, United Kingdom) have introduced new approaches into their tax systems, associated with environmental considerations. Proposals for changes aimed at lowering negative environmental pressures were referred to as "green tax reform."[2] In most cases, the reforms were undertaken with the motto of not increasing the overall tax burden. Neutrality in terms of effect on net incomes was to be achieved through limiting the existing taxes.

Three categories of undertakings can be distinguished in the comprehensive approach to ecological tax reform:

1. Decreasing the disparities in the economic system caused by taxes,
2. Shifting the structures of the existing taxes,
3. Implementating taxes with environmental background.

There are several ways of implementing the new environmental tax actions. They can be imposed on emissions of pollutants into the air or water. They also can be imposed on certain products, which has recently been more and more common. Since the beginning of the 90s, many new taxes have been imposed on products such as packaged mineral fertilizers, pesticides, batteries, solvents, oils, tires, razors, and disposable cameras. An OECD publication lists fifty-one economic instruments of this type in eleven countries, not including instruments in the power sector.[3] This does not obviate the circumstance that in many countries the number of products that are harmful to the environment is still immense, and these products are either not subject to taxation or not taxed sufficiently highly.

It should be noted, however, that efforts to implement a common EU carbon tax have failed so far. It can be taken for granted, nevertheless, that the countries leading in tax reforms will be willing still to promote this idea and to implement changes in accordance with European Union practice. Popularizing this topic in international forums will contribute to the more efficient functioning of ecological taxes. Only common adoption of similar solutions in the field of new taxes will decrease the problem that the reformed taxation weakens the competitiveness of the domestic economies where it is applied and causes additional burdens for the users of the environment.

This is the first study where the implications of ecological tax reform in Poland have been so thoroughly analyzed using general equilibrium modeling. Similar research has already been carried out in the countries of Western Europe.[4] This study focuses mainly on potential costs for the economy that may occur in the case of ecological tax reform in Poland. The year 1995 has been adopted as a benchmark because of the availability of balance sheets in the form of an input-output table for the entire national economy.

Environmental Policy Issues

Poland has an effective system of emission fees. Policy makers believe that if such a system ensures financial resources for environmental protection investment, radical modifications of taxes can be deferred. Ecological tax reform, however, should become an object of interest for Polish decision makers in the environmental protection sector. There are both external and internal reasons for this.

External motivation comes from the intensive efforts of the European Union countries to reform their own tax systems. It seems that the most environmentally advanced countries will continue to reform their tax systems in spite of EU reluctance. The Scandinavians and the Dutch are willing to "sell" their approach to the newcomers. There are signals that they would like to share their experience with accession candidate countries, very likely because that would generate more support for a European discussion regarding an increase in environmental taxes.

The second, internal basis for interest in ecological tax reform is the functioning of domestic environmental protection policy. The year 2000 was one of several subsequent years of decreasing revenue inflows from fees and fines for economic use of the environment. The institutions that manage funds for environmental protection and water management are gradually losing their significance in the structure of financing sources for environmental protection investments.

As a result, the indicator of the share of investment expenditures for environmental protection in GDP in 1999 dropped for the first time since the economic crisis of the 1980s (see Table 6.1). Moreover, the preliminary data for 2003 show a further alarming decrease: investment expenditures for environmental protection dropped to the level of around 0.6 percent of GDP. All of this happened after the close of the environmental chapter of the accession debate and before real adoption of the most difficult EU environmental standards. The total compliance costs, estimated at USD 30–50 billion, clearly show the scale of the environmental protection investment problem.

It is more and more obvious that emission fees are losing their importance in Poland. There are many reasons for this threatening breakdown, and one of them is the limited effectiveness of the old system of financing environmental protection. Technological changes, radically controlled emissions, reduced pressure on the environment, and questionable administrative reforms—all this taken together implies an unavoidable decrease in revenues from emission fees. Because of perpetual erosion of their base, emission fees have to be replaced by new instruments, namely product fees and ecological tax reforms. Product fees are, as suggested by West European experience, easier to administer while they serve fiscal

Table 6.1 Environmental Protection Investment in Poland

	1996	1997	1998	1999	2000	2001	2002
Investment (billion PLN in current prices)	6.1	7.4	9.0	8.6	6.6	6.2	5.0
Share in GDP	1.6	1.6	1.6	1.4	0.9	0.8	0.6
Share in total investment	9.4	8.1	8.0	6.8	4.9	5.1	4.6

Source: GUS (Polish Central Statistical Office), *Environmental Protection Statistics Yearbook of Poland* (Warsaw, 2004), 387.

purposes. Changes in the tax system consistent with the overall concept of environmentally oriented tax reform could be a good companion measure for product fees, achieving a double dividend in the form of better environmental quality and improvement in certain economic parameters.

Production charges, already introduced on a larger scale on 1 January 2002, and new environmentally based taxes should, in the future, take over the leading role once played by emission charges. Unfortunately, ecological tax reform is not on the agenda of public discussions, and even "green" journalists are largely unenthused about the concept. In this context, our proposal stands at the very beginning of a "long and winding road." With luck, it will be the precautionary principle, rather than any shock caused by further decay of environmental investment, that will raise interest in environmentally oriented reforms in the Polish tax system.

In the Polish economy, at present, more attention is given to modifications of the existing system of fees for economic use of the environment than to ecological tax reform. On the one hand, this can be justified by the existence of a working system of emission fees and the recent introduction of new product fees. On the other hand, this neglect stems from the fact that there are other weaknesses in the current tax system, and a project to overhaul it would require additional modifications apart from an ambitious ecological reform.

The Environmental Aspect in the Model

A simulation was made using a computable general equilibrium model (CGE); the characteristics of the model as well as their respective abbreviations are presented in the Appendix. The model has been adjusted to calculate the emissions of such air pollutants as sulfur dioxide (SO_2) and carbon dioxide (CO_2). Emissions from each sector can be generated by two sources. The first source is the use of primary inputs coming from energy sectors, e.g., fossil fuels combustion. It is assumed that these emissions are proportional to the amount of primary emission in each sector. Moreover, each sector emits air pollution directly from technological processes, independently of the amount of primary energy inputs used. In this case, emission is proportional to the volume of production in each sector.

Consumption of energy inputs by consumers constitutes an additional source of pollution emissions. These are generated only by households, as the government does not directly consume energy inputs. Household emissions are proportional to the amount of fuel used in households.

The model calculates total emissions for each pollutant, summing up the types of emissions listed above. The initial levels of emissions for 1995 are 2.4 Mt SO_2 and 334.8 Mt CO_2. It calculates the overall level of emissions based on data on emissions indicators for the individual sectors, which are introduced into the model as exogenous.

Table 6.2 Sources of Atmospheric Emissions in Poland in 1995 [%]

Pollutants	Fossil fuels combustion[1]					Industrial processes
	C	F	Fe	E[2]	other	
SO_2	82	0	9	0.01	1	8
CO_2	71	0.05	18	4	4	3

Source: O. Kiuila, "General Equilibrium Modeling of the Ecological Tax Reform Implementation in Poland," in *Ecological Tax Reform*, ed. W. Stodulski (Warsaw: Institute for Sustainable Development, 2001), 60–89.

1. See Appendix, Production Factors, 3. for a description of these categories.

2. In addition to electricity and heat energy, sector E contains production of various gases: coal gas, generator gas, etc.

The highest emissions indicators are for coal (see Table 6.2); therefore, this is the most important source of sulfur dioxide and carbon dioxide formation. In Poland, coal use is high among both producers and consumers. Should an ecological tax reform be implemented, coal will have to be replaced with other energy inputs.

The model allows simulation of various instruments of economic and environmental policy. Excise tax for fuels (t_{ex}) appears in the model as *ad valorem* tax. Product fees (t_n), in turn, which can be treated as an ecological surcharge on fuels, are treated in the model as volume based taxes. In the model, they can be imposed on the use of fuels by the specific sectors and households. They have been incorporated in the equations describing relative fuel prices, and therefore they influence production costs and the structure of demand by households.

The idea of pollution emissions fees (t_{em}) is different. In the model it is added to the price of tradable emission permits. The emissions fee is included in the function of marginal costs and the equations describing the price of fuels. The equation describing the relative price of fuels for the consumers has the following ultimate form:

$$P_h^l = \left[\frac{P^l(1 + t_{ex}^l) + t_n^l}{(1 + s^l)} + \sum_{em} s_{em,h}^l (P_{em} + t_{em}) \right](1 + t_{vat}^l)$$

where P_h^l = price for fuel l for consumer h, P^l = price for fuel l at the level of marginal costs, P_{em} = price for tradable emission permits, s^l = subsidy to fuel l, $s_{em,h}^l$ = emission indicator of fossil fuel l in household h for a given pollutant em, t_{vat}^l = value added tax on fuel l, where $l \in C, F, Fe, E$. The price equation for the producers generating indirect demand for energy fuels looks similar (only without VAT).

Instruments of economic and ecological policy in the model have indirect impact, through prices of the production inputs, on marginal costs and relative prices. In the system of equations in general equilibrium, these costs have various distributions throughout the whole economy depending on the given scenario.

Scenarios

The outcomes of the model refer to 2005 with 1995 as a base year. Scenario I constitutes the reference point for the analysis and graphic presentation of outcomes, and is considered to be the basic scenario. Calculations for the six scenarios described below led to outcomes that will be compared further regarding the direction and scale of deviations of endogenous variables.

The basic scenario (I) contains the initial assumptions + tax rates at the 1995 level. The model takes into account fee rates binding in 1995:

- t_{em} (emission fee): SO_2 = PLN (Polish zloty) 0.19/kg and CO_2 = PLN 0.0001/kg.
- t_n: fuel surcharges are equal to zero.
- Other taxes (excise tax for fuels, value added tax, and subsidies) remain unchanged.

In addition to the tax rates, initial assumptions set a number of the model's parameters:

- It assumes 5 percent increase in real prices on the world market over ten years (i.e. until 2005), stability of the Polish zloty (i.e. stable real exchange rate), and an increase of the trade balance by 1.5 percent per year.
- Increase of price elasticity for exports by 25 percent implies more competitive foreign trade in the future.
- Change in access to production inputs is one of the most important factors of the economic growth in the model: annual increase of capital amounts to 0.5 percent.
- The remaining factors of production—C, F, Fe, E, Lm—are described in the model as endogenous, therefore the rates of increase in their assets are not assumed exogenously.
- The scenario assumes an increase in the professionally active population (for manual labor by 0.9 percent annually, and for nonmanual labor by 0.7 percent annually). In this way the scenario introduces the assumption that human capital investments will stimulate work productivity and will imply increase of the initial endowment of labor at nonphysical work positions.

All of the researched scenarios II–VI (see Table 6.3) accept the initial assumptions adopted in basic Scenario I and also maintain the additional assumption that other existing taxes (excise tax for fuels, value added tax, and subsidies) remain unchanged. New tax rates are the same for all scenarios II–VI:[5]

- t_{em} (emission fee): SO_2 = PLN 0.29/kg and CO_2 = PLN 0.0004/kg;
- t_n (fuel surcharges): coal C = 7%, gas F = 4%, liquid fuels Fe = 5%, electricity and heat energy E = 0.5%.

Table 6.3 Scenarios Considered and Researched in the CGE Model

Scenarios	Distribution schemes for the revenues from new tax rates
Scenario II	Tax revenues are transferred to the state budget
Scenario III	Tax revenues are divided into two parts: (a) revenues resulting from up-to-date tax rates are transferred to the state budget (b) incomes due to the increased tax rates are refunded to the enterprises in the form of subsidies for environmental protection facilities (i.e., they increase endowment of the economy in capital), balancing the increased demand for these facilities (now we are assuming that these revenues are not transferred to the state budget but are directly devoted to environmental investment)
Scenario IV	Tax revenues are divided into two parts: (a) revenues resulting from up-to-date tax rates are transferred to the state budget (b) revenues resulting from the increased tax rates are refunded to the enterprises in the form of lower social security rates for employees (in this way labor becomes cheaper)
Scenario V	Tax revenues are divided into two parts: (a) revenues resulting from up-to-date tax rates are transferred to the state budget (b) revenues resulting from the increased tax rates are refunded to the enterprises through subsidizing capital costs (in this way capital becomes cheaper)
Scenario VI	Tax revenues are divided into two parts: (a) revenues resulting from up-to-date tax rates are transferred to the state budget (b) revenues resulting from the increased tax rates are refunded to the households in the form of subsidies to private incomes

Outcomes of the Simulation

This part of the study presents, based on the outcomes derived from the model, the possible distribution of costs of the new policy in the whole economy. The analysis will focus first on the prices of production inputs, then on their impact on production costs.

The Production Inputs Market

Figure 6.1 shows changes in the marginal costs of production inputs. The figure is presented in relative units, where the base year constitutes the reference point. The white bar on the left side for each production input shows the situation for scenario I, where tax rates are stable at the level of 1995. Further comparisons

Figure 6.1 Marginal Costs of Production Inputs

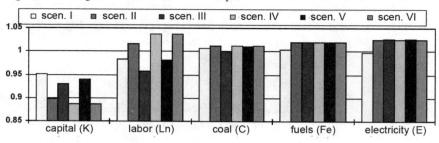

will be made relative to this scenario. The second bar for each production input shows the situation for scenario II, where the tax rates are changed, and the subsequent bars show the situations for the scenarios III, IV, V, and VI, respectively.

Due to implementation of ecological surcharges on fuels (scenarios II–VI), an increase in marginal costs of electricity *(E)* and secondary fuels *(Fe)* can be expected irrespective of the type of scenario. Capital costs (K), in turn, and costs of nonmanual labor *(Ln)* depend to a great extent on the type of scenario. A surprising outcome was achieved for coal *(C)*, because the marginal costs remain unchanged irrespective of the scenario.

Among all the inputs, only the capital costs decrease due to changes in tax rates. All the scenarios assume a future increase in the capital inputs. The increased inflow makes access to capital easier and the price of it lower. If the capital assets are not changed, the price remains unchanged, which has been verified in other simulations.

Total demand for capital does not change after implementation of new tax rates, because the amount of capital available in the market is introduced in the model as exogenous. The capital is only undergoing reallocations. This applies to all the scenarios except scenario III, where in addition to reallocation a 2 percent growth of total demand is observed.

The costs of nonmanual labor, *Ln,* like those of capital assets, can also decrease, but only in scenario III. In the market for manual labor *(Lm),* the price of this production input does not determine the change in demand because the price is an exogenous variable. Demand in this market is changing as a result of change in individual sectors, which is, in turn, a response to the change in production input prices. More expensive production inputs cannot be entirely replaced with the remaining inputs. Therefore, the volume of production in some sectors decreases, leading to a drop in demand for labor.

The most favorable impact on unemployment can be observed in scenarios IV and VI (Table 6.4). In these scenarios, a certain increase of employment and decrease of unemployment can be expected. The same conclusions apply to scenario II. Scenario III, however, may imply a negative impact on employment, because unemployment increases. In scenario V, both the labor market and the capital market remain unchanged.

Table 6.4 Change in the Number of Employed and Unemployed in Individual Scenarios as Compared to Scenario I (%)

	Employment			Unemployment		
Scenarios	Lm	Ln	total	Lm	Ln	total
II	2	0.2	1	−4	−1	−3
III	−1	0.0	−1	2	0	2
IV	3	−0.2	2	−5	1	−4
V	0	0.1	0	0	0	0
VI	3	−0.2	2	−5	1	−4

All the energy inputs are produced based on other production inputs, so their price depends on other prices. Coal (production input *C*) has the highest share in the production of the *E3* sector (electricity and heat energy), which later emerges in the market as the production input *E*.

A question arises: why does the marginal cost of coal (Figure 6.1) remain unchanged? In the long run, the producers will try to shift the cost of taxes onto buyers. If the pollution tax and the costs of pollution emissions reduction (pollution generated during coal combustion) are included in the price of coal, it turns out that the actual price of coal will significantly increase (see Figure 6.2). A dramatic increase in this price will likely result from the change in tax rates. The same is true for the price of the *Fe* fuels.

Imposing high surcharges on energy generating fuels implies that the economic agents will have to change the structure of the production inputs that they use. Such changes are subject to the parameter of elasticity of substitution. The agents give up the use of production inputs that generate high emissions of pollutants and invest in increasingly efficient technologies. Thus, the overall demand for energy-generating inputs decreases.

All the scenarios are compatible regarding the direction and scale of change in overall demand for energy-generating inputs: *C* drops by 3–5 percent, *F* by 3–6 percent, *Fe* by 3–5 percent, and *E* by 2–3 percent. The conclusion can be drawn here that decrease in the demand for energy-generating inputs will not be sig-

Figure 6.2 Coal Prices *(C)* after Taking into Account Taxes and Costs of Pollution Reduction

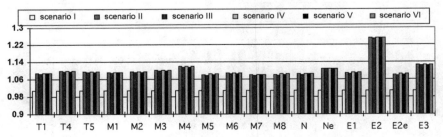

nificant under the given tax rates. It is possible that the producers will simply be able to transfer most of the tax burden to the consumers. This will be explained after the analysis of the situation in the market for goods and services.

The Market for Goods and Services

Changes in the production inputs market result in changed production costs. Figure 6.3 depicts directions of changes in production costs after the implementation of ecological tax reform. Among the seventeen sectors described in the model, only the *E2* sector (fuel and gas) is not suitable for observation because its prices are determined in an exogenous way. As compared to the benchmark equilibrium (1995), where all the marginal costs for all the sectors are set at 1, these costs for the individual sectors are, as a rule, more differentiated than for the considered scenarios.

Outcomes of the calculations suggest that new tax rates for energy-generating inputs, if set at the level assumed in the analyzed scenarios, will probably have only a slight impact on production costs. Sectors where capital plays a significant role can even benefit from the new tax policy, because their production costs will decrease.

The nonlinear programming models also allow calculation of shadow prices. Using shadow prices, we can examine the impact of price changes in the production inputs markets on the production costs. Figure 6.4 shows the impact of prices of all energy-generating inputs—*C, F, Fe, E*—on the shadow price of energy. On

Figure 6.3 Marginal Costs of Production

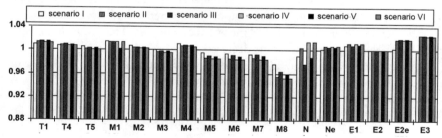

Figure 6.4 Shadow Price for Energy-generating Inputs

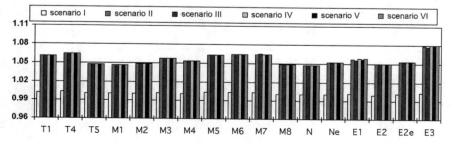

the basis of this price, we can assess the role of the prices of the individual energy-generating inputs in shaping the ultimate price for the producer in the given sector. The direction of changes in these prices is the same for all the sectors. The scale of changes also is comparable.

Large changes in the shadow price of energy have no significance for most of the sectors, with the exception of E2e (refinery industry) and E3 (electricity and heat energy production). This is due to a fall in the price of other production inputs. If we take into account shadow prices for all the production inputs at the same time (Figure 6.5), the scale and direction of changes in the majority of the sectors will be the reverse of the shadow prices of the energy inputs (Figure 6.4). The ultimate producer price incorporates, in addition, the costs related to intermediate demand.

Changes in production costs will imply price changes and a consequent shift in demand for the specific goods. A decrease in foreign demand for domestic production can be observed in the sectors where production costs increase as a result of the implementation of new tax rates, which relates especially to the electricity-generation sector *(E3)* and refinery industry *(E2e)*. In most sectors, however, foreign demand does not decrease regardless of the scenario.

Households' demand may undergo significant and not easily predictable changes. The starting point is price change for the consumers who, as the ultimate recipients of goods and services, bear the costs of VAT—in other words, all prices from the base year are increased by the rate of this tax. As a result of the new tax policy, change in production costs induces price changes. The direction and scope of these changes are similar to the changes in the production costs. Exceptions are the coal sector *(E1)* and the refining sector *(E2e)*, where the prices rise out of proportion to the increase in the production costs. Consumption of the products of the *E1* and *E2e* sectors, however, is directly related to sulfur dioxide and carbon dioxide emissions. Additional costs must be borne to reduce these emissions. The model assumes that the costs of reduction will be immediately included in prices of goods whose use results in pollution, and that they will be proportional to this use. As a consequence, the prices for the products of the coal and refinery sectors will significantly increase.

With small changes in the parameters, the consumers react forcefully to the new economic situation created with higher tax rates (Figure 6.6). Such behav-

Figure 6.5 Shadow Prices of Production Inputs (capital, labor, and energy inputs)

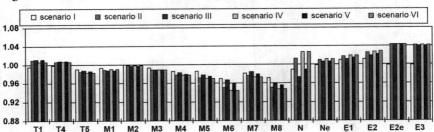

Figure 6.6 Household Demand in Billions PLN (logarithmic scale)

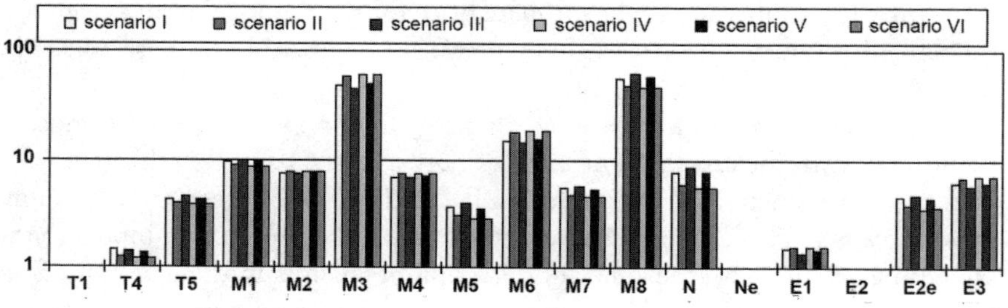

ior means that the shock of higher tax rates on goods in the energy-generating sector is of high significance to consumers, in comparison with small economic changes. In an adverse situation, in turn, if the same shock is accompanied by more significant changes in the economy, the reactions of consumers are weaker, because the economic changes are then of a higher significance. Total household demand may even increase modestly.

Changes in the production costs and in demand result in changes in the production volumes of individual sectors (Figure 6.7). In the market system, as simulated in the model, growth in some sectors is significantly hindered *(M7, M8, E1, E2e)*, while other sectors benefit from this situation *(M3 and M6)*. A surprising effect of the new tax policy is that the sector of commercial services *(M8)* may suffer at least as much as the energy-generating sectors. Outcomes of other simulations have confirmed this surprising effect induced by the demand-supply mechanism. Despite decreased production costs in the commercial services sector (Figure 6.3), production in this sector drops due to the behavior of consumers (Figure 6.6).

The coal sector *(E1)*, according to the predictions of the governmental policy, will reduce coal extraction. A slight drop in production of electricity and heat energy *(E3)* can be expected due to high costs. Scenarios II, IV and VI turned out to be the most beneficial from the point of view of production. Total production in these scenarios increases by 1–2 percent, an outcome confirmed by other simulations. Scenario V turned out to be neutral regarding change in total production, and scenario III reveals a small negative impact on the volume of production.

Figure 6.7 Production Volume in Billions PLN

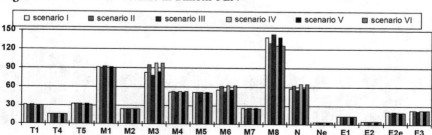

The Emission of Pollutants

The model takes into account emissions of sulfur dioxide and carbon dioxide. The higher the tax rates on fuels and emissions, the larger the impact on the reduction of pollution emissions. In the analyzed scenarios we have adopted rates lower than those recommended by OECD. Thus, emissions reductions will not be significant either (see Table 6.5). The highest impact is observed in scenarios III and V, which means that changes in the capital market have a significant influence on the level of pollution emissions.

The lower emissions in the analyzed scenarios result from decreased consumption of energy-generating inputs and decreased production in some sectors, not from installation of pollution-control equipment. None of the scenarios includes the option of emission limits, so the enterprises do not encounter this additional incentive for emission control. Poland participates in a number of international agreements on the control of emissions of pollutants. If we were to introduce such limits into the model, the drop in emissions would probably be higher.

The overall level of emissions in the economy will be strongly influenced by changes in emissions from the *E3* sector, as this is the single largest polluter in the country: 60 percent and 50 percent of total emissions of SO_2 and CO_2 respectively. This impact is the strongest in scenarios III and V.

Income Distribution in the Society

Statistical data show that rich households consume more energy than the poor ones. Poor households, in turn, are characterized by a higher share of expenditures for energy in their total expenditures, compared to rich households.

Disaggregation of households into the rich and the poor ones allows more detailed analysis of the impact of the ecological tax reform on the standard of living of these two income groups. We can both compare demand changes for individual groups and analyze beyond the model the changes in their welfare level using general concepts of welfare measurement—equivalent variation, EV, or compensating variation, CV.

According to the calculations made with the use of the CGE model, the poor households react much more energetically to any economic changes than do the rich households. Moreover, the demand of the poor consumers significantly increases in scenarios II, IV, and VI (where it more than doubles), while in scenario III it decreases. Such behavior has been confirmed in various other simulations;

Table 6.5 Total Sulfur Dioxide and Total Carbon Dioxide Emissions Relative to Scenario I [%]

Scenario	I	II	C	IV	V	VI
SO_2	100	98	94	99	96	99
CO_2	100	99	95	99	96	99

however, the scale of changes varied significantly. This problem has already been described in the explanation of Figure 6.6.

Careful interpretation of the results suggests that implementation of ecological tax reform may in fact have a progressive distribution effect in the cases of scenarios II, IV, V, and VI: the demand of the rich households for all goods decreases while the demand of the poor households increases. Such a conclusion is confirmed by the values of equivalent variation and compensating variation.

Table 6.6 presents a comparison of changes in the welfare of households. Outcomes based on CV and EV are the same, which confirms the robustness of the calculations. As can be seen in the table, ecological tax reform leads to a decrease in the welfare of the richer households and to an increase in the welfare of the poorer households, except of course in scenario III, where the situation is reversed.

Summary of the Outcomes

Simulation of the impact of ecological tax reform on the Polish economy and on income distribution, carried out using the CGE model, allows the following conclusions to be drawn:

1. The structure of pollution will change: enterprises will reduce factors that are linked to high emissions of pollutants and will invest in better and better technologies.
2. The increase in the costs of generating electricity and heat energy, which can be linked to the impact of the tax reform, is not significant.
3. The long-term impact of reaching compliance with the requirements imposed by the reform on prices and production in other sectors (not energy-generating ones) seems to be beneficial in scenarios II, IV, and VI for the food and drink industry, agriculture, and the noncommercial services sector, but it can turn out to be unfavorable for the commercial services sector; in the case of scenario III, the heaviest burden is imposed on the food and

Table 6.6 The Values Reflecting Welfare Changes (measured using CV or EV) for Specific Groups of Households

	Change in relation to Scenario I			
	Share in income (%)			(billions PLN)
Scenarios	rich	poor	total net effect	total net effect
II	−56	236	1	1.78
III	27	−100	2	4.58
IV	−70	294	0	0.73
V	−5	29	2	3.40
VI	−70	295	0	0.85

drink industry, and in scenario V none of the non–energy generating sectors suffers.
4. The real prices[6] of the majority of goods will not change, and in the case of commercial services even a decrease of real prices can be expected. Only the prices of goods generated in the sectors of electricity and heat production, crude oil and coke refining, and noncommercial services will rise.
5. The sectors that directly bear the burden of tax reform (i.e., the metal industry, the mineral extraction industry, the transport industry, the municipal services sector, the coal industry, the refinery industry, and the electricity production industry) will not break down.
6. The welfare of households likely will not decrease.
7. The burden of the new policy will be placed on the richer groups in the society, which may experience a consumption drop (except in scenario III).

The increase in prices of energy inputs leads to both direct and indirect results. The calculations prove that with every increase in energy prices the decrease in the welfare of the richer households is higher than the decrease for the poorer households. A breakdown into the types of energy inputs reveals that increased prices will cause the highest losses for all consumers in the case of petrol and other fuels generated in the *E2e* sector. Such an outcome means that subsidizing energy prices is regressive by nature: though it helps poor consumers, it is rich consumers who benefit the most from the subsidies. Thus, subsidizing energy consumption does not improve the efficiency of income distribution, while the proposed tax reform can achieve this goal.

Some conclusions can be drawn based on macroeconomic indicators. Table 6.7 shows selected economic indicators for the analyzed scenarios. The outcomes unambiguously indicate positive changes in scenarios II, IV, and VI. Thus, we can conclude that for the Polish economy, the best way to implement ecological tax reform is to implement scenario IV or VI. In the remaining scenarios, certain decreasing tendencies may arise. This confirms analyses carried out in Western European countries,[7] which suggest that the most beneficial option is to carry out ecological tax reform while decreasing the burden on the labor force or on households.

Table 6.7 Economic Indicators (not taking into account inflation)

Scenario	Gross value added	Indirect consumption	Gross output	Total supply	GDP	Unemployment
II	2%	1%	2%	1%	2%	−3%
III	−3%	−2%	−2%	−2%	−3%	2%
IV	3%	2%	3%	2%	4%	−4%
V	−1%	−1%	−1%	−1%	−3%	0%
VI	1%	1%	1%	1%	3%	−4%

Conclusions

Based on the projected outcomes, it can be stated that long-term results of the proposed tax reform should not hamper economic development, provided that the appropriate scenario is envisioned. The implemented changes may turn out to be a significant incentive for making the production processes more efficient. As an effect, higher prices of energy inputs would faster approximate the level of prices observed in the European Union countries.

The projected outcomes found in this study do not provide categorical solutions; however, they are positive and convincing enough to initiate a serious public and political debate on the conditions and potential ecological benefits of ecological tax reform in Poland. The potential benefits when compared to the present situation are as follows: a more reliable and regular source of revenue for environmental investment; removal of numerous inefficient or unsustainable tax incentives; and a more competitive economy in the future.

Appendix: Characteristics of the Model

Basic Assumptions

The model looks for solutions in accordance with the neoclassic theory of general equilibrium: it calculates the prices and volume of production, which equalize demand with supply in all markets and make marginal profits equal to zero in all sectors. In the situation of excessive supply the equilibrium price is set at zero.

In addition, the model assumes that all the sectors can be, to a certain extent, price-generating. The decision as to which sectors will be price-makers depends exclusively on price elasticity of demand for exports. Each sector as a whole can be a price-maker in the model, but the specific producers can only be price-takers.

Table 6.8 lists the most important features of the model. All the prices in the model are described as relative and for the benchmark equilibrium they are normalized to unity. It means that the model does not contain one selected price *(numeraire)*, which would be used to represent all other prices.

Classification of the Sectors

For the needs of the model, branches of the Polish economy have been aggregated into thirteen production sectors and four power sectors on the basis of economic criteria—market power, protectionism, share in international trade—and environmental criteria, i.e. the level of emission of various pollutants per unit of production (see Table. 6.9).

Production branches include five polluting sectors and eight nonpolluting sectors. Additionally, two production sectors produce goods with insignificant exports.

Table 6.8 Characteristics of the Model

Number of variables and parameters:	Approximately 300 simultaneous equations (mostly nonlinear), the same number of variables and over 200 parameters; over 1,200 values of endogenous variables
Domestic supply:	17 sectors (4 power, 3 services, 7 industry, transport, construction, agriculture); production function: combination of Leontief's function and nested CES functions
Domestic demand:	2 households (rich and poor) and government; utility function: Stone-Geary type; demand function: LES
Production inputs:	2 exogenous (capital and nonmanual labor) and 5 endogenous (manual labor, coal, other primary fuels, secondary fuels, electricity and heat energy)
Foreign trade:	Export (endogenous) and import (exogenous)
Prices:	Described as relative prices (no inflation) and normalized to the unit
Pollution (SO_2 and CO_2)	Calculated according to the sources (from fuels and industrial processes)
Instruments:	Command-and-control: emission standards; economic: emission fees, excise tax for fuels, product fees, VAT, subsidies, pollution emission fees, tradable emission permits

The power branches are divided into three polluting sectors and one nonpolluting sector.

Production Factors

Each sector uses seven production inputs to generate its own production: capital, two types of labor, and four energy sources. In addition, each sector uses the production of other sectors and its own production, which together constitute intermediate demand. The model assumes no mobility of capital and labor. The remaining production inputs are mobile. Demand for all production inputs has been established as endogenous, and supply is treated differently depending on the production input:

1. Capital (K) determines economic growth in an exogenous manner. Thus, we assume that we know the long-term balanced growth path, and for its description in the model we introduce a forecast rate of capital growth in the economy.
2. The labor market is divided into two parts: manual *(Lm)* and non-manual *(Ln)*. Supply of manual labor is set in the model in an endogenous way, and the price for its work is introduced in an exogenous way. Alternatively,

Table 6.9 Classification of the Economic Sectors in the Model

Sectors	Symbol	Sectors with high emission of pollutants	Sectors with low emission of pollutants
		(according to SO_2 and CO_2 emissions per unit of production)	
P R O D U C T I O N	T1	Metallurgy (steel works and smelters)	
	T4	Mineral industry (quarry, building materials, ceramic, glass)	
	T5		Timber and paper industry, other production
	M1		Electro-machinery industry
	M2		Light industry
	M3		Food industry
	M4	Chemical industry	
	M5		Construction
	M6		Agriculture, forestry, fishing
	M7	Transport	
	M8		Comercial services (trade, hotels, restaurants, architecture, etc.)
	Ne	Municipal services	
	N		Noncommercial services (health care, social services, education, etc.)
P O W E R S E C T O R	E1	Coal industry	
	E2e	Coke and refining industry	
	E2		Fuel and gas industry
	E3	Electricity and heating, gas production, water treatment and distribution	

Source: O. Kiuila, "General Equilibrium Modeling of the Ecological Tax Reform Implementation in Poland," in *Ecological Tax Reform,* ed. W. Stodulski (Warsaw: Institute for Sustainable Development, 2001), 60–89.

for nonmanual labor the price of work is endogenously calculated in the model and supply is treated as exogenous. Labor costs have been divided into three components: net salary (57 percent), income tax (13 percent), and insurance premiums and services for workers (30 percent).[8]

3. Sources of energy: coal *(C)*, liquid and gaseous fuels *(F)*, secondary fuels *(Fe)*, electricity and heat *(E)*. One can note that in the model the four energy inputs correspond with energy sectors. Thus, each energy sector has a double role in the model: on the one hand it identifies the production input *(C, F, Fe, E)*, and on the other hand it is simultaneously a production sector (E1, E2, E2e and E3, respectively). The supply of energy inputs is described as endogenous in the model.

Households

Total demand includes all the components of domestic demand, such as indirect demand, household consumption, and government demand. In order to assess the impact of tax policy on the distribution of incomes in the society, the model has been adjusted to reflect simulations of the distributive effect. Additionally, households have been divided into *rich* and *poor*. The average monthly income per person has been adopted as the criterion of division. This value is equal to the median for eight income groups adopted by the Central Statistical Office in Poland.[9]

Notes

1. The original research was funded by the Institute for Sustainable Development in Warsaw.
2. T. O'Riordan, ed., *Ecotaxation* (London: Earthscan, 1997).
3. OECD, *Sustainable Development: Implementing Appropriate Framework Conditions* (Paris: OECD, 2000).
4. See, among others: A. L. Bovenberg and L. H. Goulder, "Costs of Environmentally Motivated Taxes in the Presence of Other Taxes: General Equilibrium Analyses," *National Tax Journal* 50, no. 1 (1997): 59–87; K. Schlegelmilch, ed., *Green Budget Reform in European Countries* (Berlin: Springer, 1999); F. Bosello, C. Carraro, and M. Galeotti, "The Double Dividend Issues: Modeling Strategies and Empirical Findings" (paper presented at the Symposium "Environment, Energy, Economy: A Sustainable Development," Rome, October 1998); B. Bosquet, "Environmental Tax Reform: Does It Work? A Survey of the Empirical Evidence," *Ecological Economics* 34, no. 1 (2000): 19–32.
5. The level of fees has been established based on domestic and international experience.
6. Changes in nominal prices, i.e. inflation, are not observed in the model.
7. Bosquet, "Environmental Tax Reform."
8. Cost components have been calculated based on GUS (Polish Central Statistical Office), *Statistics Yearbook of the Republic of Poland* (Warsaw: GUS, 1997).
9. GUS, 1997.

Part Three

WATER POLICIES AND INSTITUTIONS

Chapter 7

THE CZECH REPUBLIC
From Environmental Crisis to Sustainability

Václav Mezřický

Sustainable Development and the Floods of 2002

The recent floods in the Czech Republic, the worst in 150 years in many places, are radically changing all previous conceptions of sustainable development and its priority. Ecological models based on the theory of the ecological footprint and environmental space only in the traditional form, i.e. the non-catastrophic form, are receding to the background.[1] The main concern is no longer a reduction in human activities to the limits of ecological carrying capacity. Coming to the fore, instead, is a concept of environmental space defined by possible recurrence of the flood disaster, perhaps more than once.[2]

Therefore, the main possible ways of limiting floods have begun to be inventoried. They include:

1. Revitalization of the land, which requires planting mixed forests, limiting the impact of irrigation and drainage, etc.,
2. Construction of new dams,
3. Construction of dry reservoirs—polders—which would be filled only during floods,
4. Construction of ponds and lakes,
5. Digging "detour" canals around towns and cities, and
6. Manufacture of portable flood walls.[3]

One also encounters another, completely opposite forecast, according to which we may expect a period of perhaps fifteen years of dry weather. Both cases depend on processes that cannot be influenced by a single state, and any solution must necessarily be of a global nature. However, in the last analysis, any such solution must begin with national or regional strategies for sustainable development.

From Environmental Protection to a Strategy for Sustainable Development

Until the early 1980s, the totalitarian communist regime devoted practically no attention to the environment. All activities of various official and semiofficial organizations and civic groups concerned with this issue were continually monitored by the political police. The subject of the environment and global ecological problems was considered incompatible with the vision of a communist society. This was logical, in part because environmental protection and management of natural resources required restrictions on the idea of unlimited development, which was an intrinsic pillar of the reigning Marxist ideology. The nature of environmental awareness on the part of the broad public also corresponded to this situation: although the population perceived individual local problems that people experienced directly, it had no global overview and, especially, knew almost nothing about the causal relations of the phenomena experienced.

Inventory of the State of the Environment: Plans for Ecological Policy

The breakthrough to a new stage was accomplished by the first inventory report on the state of the environment—*An Analysis of the Ecological Situation in the Czechoslovak Socialist Republic.* It was prepared in 1983 by the Ecological Section of the Czechoslovak Biological Society, associated with the Czechoslovak Academy of Sciences.[4]

The report showed, for example, that in production of emissions per square kilometer the Czechoslovak Socialist Republic (at that time a unified state of Czechs and Slovaks) ranked second place among the countries of Europe, ahead of such countries as the Federal Republic of Germany and the United Kingdom. About 3 to 4 million citizens (of a total population of 15 million) were living under the influence of emissions, and emissions had damaged about 40–60 percent of all forests, which covered 30 percent of the total land area of the country.

The report still failed to express precisely the degree of water pollution, because there were no figures available on this at the time. It only pointed out the rising trend of pollution as a result of restricted construction of sewage treatment plants—for example, that between 1970 and 1980 pollution had increased by 70 percent.

As regards land, the report referred mainly to a deterioration in the quality of soils as a result of erosion, acidification, and use of excessive quantities of chemical fertilizers, pesticides, and heavy machinery. Although large, protected nature reserves at that time measured about 14,680 square kilometers, i.e. 11.4 percent of the country, a total of 75 percent of these areas had been tainted and 25 percent seriously damaged. The consequences for free-ranging animals were catastrophic: under threat were 30 percent of the fish, 60 percent of amphibians, 30 percent of reptiles, 30 percent of birds, and 35 percent of mammals.

Waste production could only be estimated based on the quantity of mineral raw materials mined, because data on wastes were not collected at that time. The report reached the conclusion that about 35 metric tons of wastes per capita were produced each year. It remains to add that reports prepared later by the totalitarian regime did nothing to change these basic data, nor did new and more thorough inventories of the state of the environment made after 1989.

The authorities of the totalitarian state suppressed the report immediately after it was completed. Nevertheless, via dissidents it made its way abroad. There it was widely published, for example in the French newspaper *Le monde,* and through broadcasts on the Voice of America and Radio Free Europe the Czechoslovak public, too, became acquainted with it. From this time on we can speak of environmental awareness on the part of the broad Czechoslovak public. This played one of the decisive parts six years later in the "Velvet Revolution" that led to the establishment of a democratic regime after November 1989. The broad publicity received by the *Analysis of the Ecological Situation in the Czechoslovak Socialist Republic* of 1983 finally forced the totalitarian regime to devote attention to the environment, without, however, admitting that the causes of its critical state were mainly of a systemic nature.

From a historical point of view, it is indisputable that the exploitation of the country's territory beginning late in the nineteenth century exceeded ecological carrying capacity, especially in the Czech lands. Around 1900, before the fall of the monarchy, 75 percent of the industrial potential of the Austro-Hungarian empire was located in Bohemia and Moravia. Thus, between the two world wars, Czechoslovakia was among the ten most industrialized countries of the world, partly as a result of development of the arms industry in Slovakia.[5]

After the totalitarian regime seized power in 1948, industrialization continued at a dizzying pace. By 1986, the volume of industrial production and construction had increased roughly thirteen fold, which means that the volume for the whole year of 1948 was now produced in less than a month. The totalitarian regime itself began to admit that most economies in developed industrial countries got by with half of Czechoslovakia's annual per capita production of steel, which was 970 kilograms at the time.[6]

Besides the structure of the economy, additional factors contributing to the catastrophic state of the environment were the poor quality or even lack of relevant laws, the ineffectiveness of economic instruments, the system of socialistic ownership, and the system of "planned management" of the economy. It suffices to note only briefly that, for example, there was no law on wastes or chemicals, and that the law on the protection of the atmosphere was very primitive. In addition, given the uniform system of state ownership, economic stimuli could not effect a change in the behavior of economic entities—i.e. of individual manufacturing enterprises—because they had no independent economic status.

In the new conditions after 1989, a new, more thorough inventory of the state of the environment was made. A "Blue Book" was issued to monitor the state of the environment in the Czech Republic. Slovakia now conducted such monitoring independently. There was also the "Rainbow Book," which indicated the main directions and most important forms of ecological policy.[7] Over the course of years these initial impulses developed into the *Státní ekologická politika* (National Environmental Policy).[8]

The functioning of market mechanisms, budget outlays for environmental protection, a new system of environmental law, and the functioning of instruments of economic stimulation brought about a radical change in the state of the environment in many respects. Pollution of the air by SO_2 (sulfur dioxide) had declined by 86 percent as of 2001, and NO_X (nitrogen oxides) by 47 percent, although this pollution is still higher than in European Union countries. Water purity also improved greatly—in the case of biological oxygen demand by 84.9 percent and in chemical oxygen demand by 77.8 percent. However, land erosion continues. More than half of all soils are continually threatened by water erosion and about 30 percent by wind erosion. As of yet, we can only estimate what losses have resulted from past and present flooding.[9] An important change is that per capita energy consumption has declined by 20 percent, and energy consumption in relation to GDP by 58 percent. On the other hand, there has been unfavorable growth in automobile transportation and a decline in the number of people transported by rail. The decline in the volume of industrial production resulting from reduced production by the weapons, steel, and machinery industries may also be regarded as positive from the standpoint of the vision of sustainable development, though not from the standpoint of social impacts in the form of unemployment.

Nevertheless, the *National Environmental Policy* and other plans prepared since 1989 remain only defensively conceived collections of measures for protection of the environment and its individual components (water, land, air, and natural areas) and, in the best cases, also measures for safe waste management. The *National Environmental Policy* was not and is not a plan for establishing relations of sustainable development. A mere suggestion of such an approach is found in Chapter VI—"The Goals of the Czech Republic's State Ecological Policy"— and in measures proposed for various sector policies, although even here we find only general ideas, which the ministries for the relevant sectors have not adopted.

A Strategy for Sustainable Development

Thus, until 2001, the Czech Republic had no strategy for sustainable development comparable to those prepared by various nongovernmental organizations in many countries of Europe during the 1990s, of which typical examples are *Sustainable Austria* and *Sustainable Germany*.[10]

Between 1998 and 2001, the first such study was prepared under the coordination of the Center for Environmental Issues of Charles University in Prague. Other studies also were elaborated in the Ministry of Environment and the Faculty of Social Studies of Charles University, Prague.[11] However, these documents again do not provide what one expects from studies of this type.[12] What the research lacks is, above all, a clearly defined methodological basis. Today, such a basis is offered by the concept of the ecological footprint and of environmental space, despite various problems and possible reservations. Only on this basis can one successfully derive the postulates for limiting and changing human activities with the goal of establishing relations of sustainable development.[13]

In light of this fundamental deficiency and in comparison with *Sustainable Austria* and *Sustainable Germany*, the Czech strategy is not sufficiently concrete, and its structure of priorities does not correspond to their actual importance in human activities (e.g., excessive emphasis on tourism). The concentration on global problems, rather than on those connections and relations of sustainable development within the central European region that have direct significance for the national conception of sustainability, is exaggerated and nonfunctional.

Prospects for Establishing Sustainable Development in the Czech Republic

People are coming to realize that every strategy for sustainable development must first come to terms with some fundamental theoretical issues and must define its basic starting points. As stated above, this starting point is the concept of the ecological footprint and environmental space.

According to some preliminary calculations, with an ecological footprint of 4.2 hectares per capita no more than four million people should live in the Czech Republic. Although this is in any case only an approximate figure, undoubtedly it should influence discussions about usable environmental space. In any case, it is a permanent *memento* both for decision-makers and for all other stakeholders. Comparison of ecological footprints of neighboring countries of the region can play a similar role as the ecological footprint of the Czech Republic (see Table 7.1).

The strategy should also follow the data concerning five key consumption categories, i.e., food, housing, transportation, goods, and services. The data on the level of the individual's particular ecological footprint, taking into account the consumption in each category, may represent a challenge to change the consumption pattern of an individual.

A huge amount of money has already been spent to elaborate the strategies. Therefore, we may expect that the task of finally working up the strategy will be taken up by the only institution that can marshal the additional financial resources, the Ministry of Environment. It is more than probable that it will invite certain NGOs and certain representatives of the academic sector to participate in the work. NGOs have already been involved in previous strategic projects, but because they have never been leaders of such projects they have had limited opportunities

Table 7.1 Ecological Footprints of the Czech Republic and Neighboring Countries

Country	Population (million inhabitants)	Ecological footprint	Ecological deficit
Czech Republic	10.263	3.9	−1.4
Hungary	10.454	3.1	−0.5
Germany	81.594	4.6	−2.8
Poland	38.557	3.9	−1.9
Austria	8.045	4.6	−0.5

An *ecological footprint* is the area of ecologically productive land that an individual person, community, city or country, or a particular branch of industry or agriculture, requires for securing its consumption of resources and disposal of its wastes. *Environmental space* is defined in an opposite way. It is the maximum quantity of natural resources that a human (or humanity) can use *sustainably* and without violating the global natural balance.

for real participation. What form the strategy should take, i.e., whether it should be a single document or a set of various plans, for example for energy, transportation, agriculture, etc., remains unclear.

A comprehensive consideration of the political expectations that may be connected with the elaboration of the Strategy of Sustainable Development is simple: the calculation of the ecological footprint itself, which is the result of human activities in confrontation with an adequate environmental space, represents an existential ethical challenge. The public could recognize that by the overexploitation of the environmental space we limit our fellow citizens, other nations, and future generations, or it may even recognize that the overexploitation is suicidal. In this case, the following effort to reduce the ecological footprint would be a logical consequence of the existentially reflected change of values.

If the public will not admit the existential ethical importance of the ecological footprint, then it will be dismissing the calculated ecological footprint as something of no more importance than reports of football club results. It is, in fact, an issue of the most basic human freedom, including the freedom to commit suicide.

The creation of the strategy, however, *is not* and *cannot be* a political event that would immediately change the orientation of the political scene. Neither the Austrian nor the German policies manifested any significant changes after NGOs published national strategies for sustainable development. Nevertheless, strategies play a permanent role in policies. They influence governments' political approaches toward issues such as climate change and nuclear energy, and they indicate the consequences of actions, for instance agricultural and forest soil withdrawals. Strategies in connection with the activities of NGOs, green parties (in Germany, Greens are members of the government), and the informed public become a relatively clearly defined challenge that can step by step affect a broad spectrum of stakeholders.

However, every attempt to establish relations of sustainable development begins with an individual ethical decision. In the Czech Republic, this means overcoming internal value barriers and stereotypes such as, for example, reliance on the power of science and applied technology. For now, we can only guess how the floods of 2002 might change these attitudes.

Finally, we must not overlook the above-mentioned need for international cooperation, at least among neighboring states within a region, surely. As early as 1997, the Institute for Environmental Policy attempted to find a common basis for cooperation among the states of central Europe, especially in the areas of energy and transportation.[14] This attempt should not be the end of the story.

In summary, it can be said that steadily increasing attention is now being devoted to conceiving strategies for sustainable development. However, so far there is no very clear idea as to the optimal form of such statements. This conference may serve as an impulse for the participating countries to pursue discussions of this topic, and possibly to attempt to conceive, together, the basic structures of sustainable development.

Postscript

Subsequent to the submission of this essay, the government of the Czech Republic took steps toward the creation of a strategy for sustainable development. The Government Council for Sustainable Development was created on July 30, 2003 (Government Resolution No. 778). The Council then supervised the preparation of the *Czech Republic Strategy for Sustainable Development* which was adopted by the Government on December 8, 2004 (Resolution No. 1242).

Notes

1. Mathis Wackernagel and William Rees, *Our Ecological Footprint: Reducing Human Impact on the Earth* (Philadelphia: New Society Publishers, 1996).
2. "Proč se počasí zbláznilo" (Why has the weather gone crazy?), *Věda* (Science), supplement to *Lidové noviny* (The People's News), 17 August 2002.
3. Ibid.
4. Ecological Section of the Czechoslovak Biological Society, *Rozbor ekologické situace v ČSSR* (An analysis of the ecological situation in the Czechoslovak Socialist Republic), a study produced in association with the Czechoslovak Academy of Sciences (Prague, 1983).
5. "Československo o sedmdesát let později, situační zpráva" (Czechoslovakia seventy years later: a situational report), *Svědectví* (Paris) no. 85/86 (1989).
6. "Zásady státní koncepce tvorby a ochrany životního prostředí" (Principles of a state plan for creation and protection of the environment), supplement to *Hospodářské noviny* (Business News) 20 (1986).
7. Ministry of the Environment, *Životní prostředí České republiky* (The environment in the Czech Republic) (Prague: Academia, 1990).
8. Ministry of the Environment, *Státní ekologická politika* (National environmental policy) (Prague: Ministry of Environment, 2001).
9. Ústav pro ekopolitiku (Institute for Environmental Policy), *Proměny politiky* (Transformations in policy) (Prague: Institute for Environmental Policy, 2001).

10. Michael Kosz, ed., *Action Plan "Sustainable Austria," Eine Untersuchung im Auftrag von Friends of the Earth Österreich* (Vienna: Friends of the Earth, 1994); Bund und Misereor, *Sustainable Germany* (Wuppertal: Wuppertal Institute for Climate, Environment and Energy, 1995).
11. M. Potůček et al., *Průvodce krajinou priorit pro Českou republicku* (A guide to priorities landscape for the Czech Republic) (Prague: Centrum pro sociální a ekonomické strategie, Universita Karlova v Praze, Fakulta sociálních věd, 2002).
12. *Národní strategie udržitelného rozvoje České republiky, Projekt koordinovaný Centrem pro otázky životního prostředí University Karlovy* (A national strategy for sustainable development in the Czech Republic: A project coordinated by the Charles University Center for Environmental Issues) (Prague: Charles University Center for Environmental Issues, 1998–2001).
13. Wackernagel, *Our Ecological Footprint;* V. Třebický, "Ekologická stopa" (The ecological footprint), in *Unese Země civilizaci?* (Will the Earth kidnap civilization?) (Prague: Ministry of the Environment, 2000).
14. Institute for Environmental Policy, *Sustainable Development for Central Europe* (Prague: Institute for Environmental Policy, 1997).

Chapter 8

THE TISZA/TISA TRANSBOUNDARY ENVIRONMENTAL DISASTER

An Opportunity for Institutional Learning

Jim Perry, Eszter Gulácsy, and László Pintér

Environmental management decisions require the integration of technical knowledge with the social and cultural milieu surrounding the individuals making the decisions (i.e., institutional as well as local and national cultures). In Central and Eastern Europe (CEE) and the Former Soviet Union (FSU), environmental decision makers are being asked to play increasingly complex roles. As individuals, most of their background and training is still based in a centrally planned and centrally directed society. Today they are being asked to make decisions on a more participatory and open basis. When such decisions deal with a multi-national or transboundary condition, that cultural complexity increases. Upon occasion, our environmental management actions are not successful, resulting in negative impacts. Those pollution incidents and the associated environmental problems have many causal roots. These include technical and technological factors (e.g., inadequate scientific knowledge, inability to adequately predict future events, or inadequate infrastructure), and the vulnerability of ecosystems, some of which are more sensitive to certain impacts than others. Most environmental management decision makers are familiar with and attempt to compensate for those sources of uncertainty. However, the influence of the individual is significantly underaddressed in discussions of environmental decision making.

In this paper, we use a mining and water pollution disaster—the cyanide spill in the Tisza/Tisa[1] River from Romania to Hungary—to investigate those relationships. The Aurul processing plant in Romania uses cyanide to extract gold from the ore. Tailings, which are rich in cyanide and heavy metals, are stored in tailings ponds. In January 2000, rainfall and snowmelt caused the tailings dam to break, releasing more than 100,000 cubic meters of wastewater into the river and downstream into Hungary. Numerous fish and other aquatic organisms were killed, human health was threatened, and both domestic and endangered animal species were killed by feeding on the carrion. The incident was a public relations disaster, with poor communication causing conflict between the Romanian and Hungarian governments, as well as between the government and environmental NGOs in both countries. It also catalyzed the emergence of strong opposition to other mines in the Tisza watershed that rely on cyanide technology, such as the controversial Rosia Montana goldmine developed by Gabriel Resources Inc. of Canada.

We use the Tisza/Tisa spill to argue that environmental decision making is a highly personal endeavor. "The mine management," "the Romanian government," and "the Hungarian government" are not actors who make decisions. Rather, each of those consists of individuals making discrete decisions, representing a variety of interests, and operating within their own personal and institutional cultures. We argue that it is not possible to understand the causes of environmental management problems, nor is it possible to develop adequate responses to such disasters, without considering the role, capacities, and interests of the key decision makers involved with the issues at hand. In the context of transboundary problems, these roles and interests are especially significant, often transparent, and significantly underaddressed. Indeed, cross-border cooperation and communication is a prerequisite of effective problem-solving and prevention of these transboundary issues. As we strive to improve environmental decision making in CEE and the FSU, and especially as we encounter transboundary decisions in these nations in transition, we must carefully consider the role of the individual decision maker when proposing technical or policy-based solutions. Strategies to strengthen any aspect of decision making (e.g., to increase its participatory nature) must address this personal issue explicitly.

Introduction

Transboundary environmental problems are embedded in the context of interacting ecological, technological, sociopolitical, and cultural forces that collectively influence the way events unfold. The context always changes, sometimes quickly and radically, as in the late 1980s, during the breakup of the centrally planned economies of Central and Eastern Europe. As the CEE case illustrates, disturbance and subsequent reorganization do not necessarily lead to unambiguous outcomes. Many elements of the pre-disturbance system (e.g., ethnic tensions predating the socialist era) survive and amalgamate with the new system.[2] This should not be surprising; technological and cultural evolution tends to follow rather different patterns and dynamics. While some aspects of the cyanide spill that occurred in early 2000 at the Aurul S. A. mine in Baia Mare, Romania, could be linked to pre-1990 history and culture, others already are products of an entirely new context that emerged in the wake of political and economic transformation during the last decade. Separation of these eras, however, is artificial; together, they form the reality of Romania, Hungary, and more broadly, the CEE and FSU region, and they collectively help shape the choices made by individual decision makers.

Before 1990, in the centrally planned economies of Central and Eastern Europe and the Soviet Union, transboundary water pollution and its consequences were chronically neglected. When serious pollution incidents occurred, they were reported as a fact of life, if reported at all, by-and-large reflecting an underlying belief that this was part of the price of progress society should be prepared to pay. But it could also reflect many other things. From the perspective of basic science, it could reflect insufficient understanding of the ecotoxicology of harmful substances, or at least a lack of interest in relevant expert knowledge by the political establishment. From the economic perspective, disregard for pollution could be influenced by the fact that many environmental parameters were (and continue to be) external to the national accounting system; therefore, costly measures to sustain them were difficult to justify. Even if economic, ecological, or human health-related costs were high, maintaining the facade of ideological supremacy and social stability required denial that environmental problems existed. This was particularly important in the context of cross-border environmental problems, where (displaying) unity and solidarity with other countries in the Eastern Bloc or among republics within the Soviet Union was not to be compromised.

While the influence of the former system prevailed in ways that helped shape the responses to the Romanian cyanide spill (e.g., the absence of ecological parameters in national accounting), the spill brought to the surface other issues that clearly were products of a market economy. The corporation responsible for the spill is a joint venture of the Romanian state, an Australian company, and local individuals. Among the financiers of the operation are major international investment houses. Compared with previous standards, the technology applied in this mine was supposed to meet both economic and eco-efficiency criteria, in addition to creating jobs in a region with chronically high unemployment. Unlike earlier, somewhat similar cases, the Tisza/Tisa disaster created significant publicity that went well beyond the region. This can be explained only partly by the seriousness of the accident. By 2000, the interest of the public, NGOs, political decision makers, and in a sense companies and investors had been primed by a series of high-profile spills. These had not only brought technical, ecological, and human health issues to the fore, but also highlighted the inadequacy of national and international legal and institutional frameworks. More than ever, they also had elevated the questions of collective and individual liability.

Chronology of the Accident

During the night of 30 January 2000, over 100,000 cubic meters of wastewater with high concentrations of cyanide and heavy metals, primarily copper, spilled into the Sasar and Lapus rivers near the city of Baia Mare in northern Romania. The spill originated at the Aurul S. A. mining site, where cyanide was used to extract gold and silver. Cyanide-laden wastewater was stored in a tailings pond. Heavy rains during the weeks preceding the accident and thawing snow in Baia Mare led to an uncontrolled rise of the pond level, resulting in an overflow of the dam.

The cyanide plume initially traveled 1.2 km to the Lapus River, a tributary of the larger Szamos/Somes River (figure 8.1). The general manager of the Romanian Environment Ministry reported that cyanide levels 700 times the norm were recorded in nearby river water after the spill. The plume traveled from the Hun-

Figure 8.1 Cyanide Concentrations in the Tisza/Tisa River in the Wake of the Baia Mare Disaster

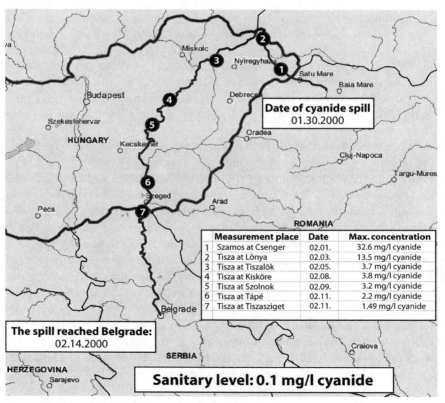

Source: Data from the Environmental Inspectorate of the Ministry of Environment and Water, Hungary; graphics by IISD (International Institute for Sustainable Development).

garian stretch of the Szamos/Somes to the Tisza/Tisa River, Hungary's second largest waterway. Extreme weather conditions in Hungary had iced over the rivers and lowered water levels, so the cyanide was not diluted as quickly as it might otherwise have been. Along with the cyanide itself, heavy metal concentrations in the water increased significantly. Hungarian authorities reported that copper, zinc, and lead concentrations all exceeded the Hungarian "heavily polluted" threshold. The plume required 12 days to pass through Hungary, from which it flowed into a stretch of the Tisza/Tisa in Serbia and then into the Danube north of Belgrade.

Was this spill unique?

The Baia Mare incident is not unusual in the CEE mining industry. On 10 March 2000, a second spill of minerals and heavy metals entered the Vaser River. The dam at Baia Borsa, owned fully by Novat, a local branch of the state coordinating company Remin S. A., ruptured due to torrential rains and snowmelt. This second spill contained aluminum, copper, zinc, and lead and threatened the integrity of the upper part of the Tisza/Tisa River, the only part of the river not heavily affected by the first spill. Most people had hoped that the upper part of the Tisza/Tisa would serve as the main source of biological recolonization for the stricken lower area.

On the night of 14 March 2000, a third incident occurred in the same watershed. In this case, the source of the pollutants was again the mine and settling pond at Baia Borsa. This dam failure of approximately three hours further contaminated the waterways, mainly with lead, zinc, copper, and aluminum.

Consequences of the cyanide spill at Baia Mare

In the Tisza/Tisa River spill, phytoplankton and zooplankton were completely eliminated when the plume passed; fish were killed in the plume or immediately thereafter. It is estimated that more than one hundred thousand tons of fish were killed along with several otters and other wild and domestic animals that fed on the dead fish or drank the contaminated water. Of the sixty-two species of fish in the Tisza/Tisa, twenty are endangered or protected; many individuals and perhaps entire populations of those fish were killed. A rare and endangered white-tailed eagle died after eating cyanide-contaminated fish.[3]

The plume contaminated the drinking water supplies of several cities and villages in the region. At its worst, the upper part of the Tisza/Tisa was contaminated with cyanide concentrations up to 100 times the limit value for drinking water. Therefore, villages and towns had to be supplied with drinking water from artesian wells, or water was trucked in and distributed in plastic bags.

The accident highlighted the lack of awareness of ecological threats or concern for them on the national level in both Hungary and Romania and also on the international level. Liability is one of the key underaddressed issues. Based on the legal framework in place at the time of the accident, the financial liability of Aurul S. A. was limited to USD 170.[4] Although the Hungarian government initiated a lawsuit against Esmeralda Exploration Ltd., Aurul S. A.'s main parent company in Australia, Esmeralda entered receivership and passed the economic burden of the spill onto the taxpayers of Hungary and Romania. This in turn points to the lack of regulation and an inadequate authorization process in Romania. However, Romania, strictly speaking, had limited interest in maintaining adequate standards of regulation and preventing harm to other states because of a gap in binding international legislation governing liability in transboundary industrial accidents. Romania was legally protected from downstream damages. It became evident that even though 96 percent of its rivers originate abroad, Hungary was politically and technically unprepared to manage transboundary pollution of its rivers.

Cyanide and Heavy Metals in Mining

Cyanide leaching is today one of the preferred methods for processing gold ores around the world.[5] Although the risks associated with the use of cyanide are well known and much criticized, in the case of the Tisza/Tisa spill one cannot blame old and worn-out equipment or obsolete technologies. Nor was the Baia Mare incident the first or most severe mining accident in recent history. In the past decade, similar accidents have happened in gold mines in the U.S., Ecuador, South Africa, Guyana, the Philippines, and Spain.[6] In gold and silver extraction, cyanide is the active agent in heap-leaching, where a dilute cyanide solution is sprayed on piles of crushed ore and the cyanide attaches to minute particles of gold to form a water-soluble compound, from which the gold can later be recovered.[7] Similarly, in the extraction of nonprecious metals such as copper, nickel, cobalt, and molybdenum, cyanide is used in the milling and concentration processes to separate metal products from the wastes.

Cyanide is highly toxic to fish and aquatic organisms. Most fish are nearly one thousand times more sensitive than are birds or mammals: whereas birds and mammals are killed by cyanide in the milligram per liter range, fish die from concentrations as low as a few micrograms per liter. The toxicity of cyanide compounds to humans is relatively high and widely known. Cyanide itself is quickly oxidized upon contact with the atmosphere and, thus, does not persist in the environment. However, cyanide is an excellent ligand, and the aqueous complexes with heavy metals that it readily forms persist in the environment for long periods of time, even after the cyanide itself has decomposed.[8] In other words, when cyanide solution is sprayed onto a heap of ore, it is not only gold but also other heavy metals that are dissolved. When this solution of cyanide and heavy metals is discharged into the environment, the cyanide quickly decomposes, but the metals stay behind, causing chronic environmental problems.

Heavy metals, often associated with cyanide, are chronic rather than acute contaminants. Metals such as lead, cadmium, and copper become buried in the sediment, and then can be mobilized by chemical or microbiological processes. This means that even though the concentration of heavy metals may be small in the water, those toxicants accumulate and may magnify in the tissues of a higher organism during its lifespan. Furthermore, some heavy metals biomagnify up the food chain; top predators (e.g., pike, raptors, humans) accumulate the highest concentrations. Since many fish species in the Tisza/Tisa are caught and distributed commercially in the region, there was a realistic threat of human exposure to heavy metals through consumption of contaminated fish. Also, heavy metals can eventually seep into the groundwater, so there is expected to be a long-term threat to agriculture as crops are exposed to those metals in water and soil. Finally, heavy metals can be responsible for very serious sublethal conditions in humans, among them kidney damage, hearing and learning disabilities in children, and delays in babies' mental development.[9]

Possible Causes of the Cyanide Spill

A chemical spill, as the accident at Baia Mare has shown, can turn relatively unimpacted ecosystems into severely polluted, potentially hazardous sites in the course of a few days. The Tisza/Tisa accident represents a valuable opportunity to use the spill as a model, analyzing the institutional and individual precursors and responses. That analysis allows us to understand more fully the underlying causes of the accident, to learn why such a disastrous spill could occur, and increase our ability to minimize the occurrence of such accidents in the future.

It is important to note that there was no one single cause for the spill. Rather, it was a product of simultaneous negligence by many actors, who collectively failed to contain the spill and, in fact, aggravated the problem once the accident occurred. Gaps in national legislation failed to establish a transparent authorization process, which could have ensured that the mining plant would be safe before it was allowed to operate. Technical causes, such as flaws in the design and management of the plant, contributed to the accident, as did extreme, but not unprecedented, weather conditions.[10]

In the immediate aftermath of the accident, there was little or no transparency in communication with the public in either country. The level of public knowledge and understanding of risks inherent in mining and related industrial processes was very low.[11] The accident was reported to the public relatively late, and the media gave contradictory accounts of how the accident happened. In Hungary, the accident appears to have been handled inadequately by individuals in the responsible ministry; reportedly, they failed to respond promptly and did not communicate with Parliament following accepted protocols.

There also are many gaps in our scientific knowledge about long-term effects of cyanide and its several breakdown compounds, which may be toxic to aquatic organisms and may persist in the environment for a long time. This was an exacerbating influence because, since breakdown compounds of cyanide are not well understood, no data on them were collected either routinely or during spill response. Finally, the lack of international agreements governing liability meant that in the Tisza/Tisa case, a company operating in one state and causing harm to another state could not be held responsible for its actions.

Learning from a Transboundary Disaster

Accidents such as the one at Baia Mare have been relatively common over the last fifteen years. We find it surprising that such a history (e.g., most dramatically the Sandoz industrial spill in Switzerland in 1986) has not served as a more useful model for international policy instruments that would guide responses to such events. For example, in 1992, the Alamosa River in the U.S. state of Colorado was severely contaminated by a burst wastewater dam; aquatic life was essentially eliminated through a 25-km stretch of the river. In Ecuador in 1993 and in South Africa in 1994, sludge and rubble from mining accidents killed twenty-four and seventeen people, respectively. In 1995, cyanide from a mine in Guyana was responsible for extensive loss of aquatic life in the Essequibo River, while in 1996, three million tons of sludge from a copper mine was released into the Boac River in the Philippines.[12] In the U.S., failure of a leach pad structure at the Gold Quarry mine in Nevada released nearly one million liters of cyanide-laden wastes into two creeks in 1997. On 29 May 1998, six to seven tons of cyanide-laden tailings spilled into Whitewood Creek in the Black Hills of South Dakota, resulting in a substantial fish kill.[13] Also in 1998, dams around a tailings pond broke at the Aznalcóllar mine in Spain, releasing three million cubic meters of sludge containing zinc, silver, lead, and copper concentrates and four million cubic meters of acidic waters into the environment and impacting more than 4,500 hectares of the adjacent Coto Doñana National Park.[14] The incident during 2000 that received the greatest public attention was the individual accident at Baia Mare, but there were several others. Just a few months after the Tisza/Tisa River incident, Sweden's worst mining waste dam collapse occurred on the Vassara River near Gallivare near the Arctic Circle, releasing a million cubic meters of wastewater contaminated with copper. The Swedish mine was owned by the same company that was responsible for the devastation of the Doñana wetland in Spain.[15]

Thus, the repeated occurrence of mining-related accidents is not restricted to one part of the world. It is not just a problem of Eastern Europe or the Former Soviet Union or developing countries, but is a worldwide issue. The question of prevention—How could this have been avoided?—was raised in connection with both the Aznalcóllar and the Baia Mare accidents. In the latter case, extreme weather conditions were cited as a reason for the accident being unforeseeable.[16] The question facing local and international society, paraphrased for the Tisza/Tisa case, is: Are 0.3–0.6 grams of gold per ton of tailings[17] plus the ancillary benefits arising from local employment and tax revenues adequate societal gain to compensate for the risk of losses in habitat and biodiversity, as well as the direct risks to public health? This question is broadly applicable; evidence from these nearly annual yet reportedly "unforeseeable and unpreventable" accidents suggests that the risk is high indeed.

The accident at Baia Mare also was similar to other transboundary events in the ways that it identified gaps and inefficiencies in existing environmental legislation. Following the industrial accident at the Sandoz factory near Basel in Switzerland, the affected governments moved together to save the Rhine, a river that had previously been considered the "sewer of Europe."[18] That accident happened in November 1986, when a fire broke out in a chemical factory. As firefighters extinguished the flames, water mixed with toxic pesticides flowed into the Rhine. The runoff killed tons of eels and other animals, and prompted a drinking water alert for 50 million people, some as far away as Amsterdam.[19] The International Commission for the Protection of the Rhine developed a program in 1987 to decrease discharge of noxious substances into the river by 50 percent, to establish a riverwide alert system, to adopt rigid safety precautions to prevent dumping of toxins, and to restore the Rhine's original flora and fauna by the end of the decade.[20] Further, the accident prompted amendment of the EEC's Seveso Directive to include storage of dangerous substances.[21] It also led eventually to the creation of the Seveso II Directive, which included a revised and extended scope; introduction of new requirements related to safety management systems, emergency planning, and land-use planning; and reinforcement of provisions on inspections to be carried out by member states.[22] As a result of the concerted effort of the countries of Western Europe, water quality in the Rhine improved significantly. In 1995, for the first time in 50 years, salmon and sea trout returned to the upper region of the river.[23]

Yet, that institutional learning apparently did not transfer to Romania or Hungary. Just a few years following the Baia Mare disaster, new controversy arose related to a new open pit goldmine in the Tisza/Tisa watershed that uses similar cyanide leaching technology.[24] Over its projected lifetime of seventeen years, the Verespatak/Rosia Montana mine, promoted by the Canadian junior mining company Gabriel Resources Inc., would lead to a 400-hectare tailings facility and 250 million tons of tailing materials. Due to concerns about the risk to ecosystems and the high social cost (including the relocation of 900 families), opposition to the project has become widespread, both in Romania and internationally. Concerns about risk and the perceived inability or unwillingness of Romanian authorities to enforce strict environmental management rules in the context of the Verespatak/Rosia Montana project have become critical factors in Romania's EU accession negotiations.

The Tisza/Tisa incident makes clear that there is much to improve at all levels if society is to be able to prevent similar accidents from occurring. On the level of plant design and management in Baia Mare, no measures regarding how the plant should operate under "uncommon but not unrealistic" weather conditions were in place. A combination of high precipitation and low temperatures during the months preceding the accident made safe operation of the plant impossible. There were no specific provisions for avoiding the overflow of the storage pond in case of heavy precipitation. The initial Environmental Impact Study carried out on the AURUL project stated that "the danger of the dam overflowing the embankment ... in the event of heavy rainfalls is out of the question."[25] Some have concluded that the regulatory authorities should never have approved the closed-circuit technology used in the Baia Mare plant.[26] However, environmental regulatory processes are complex; it is unclear exactly which authority(ies) bore responsibility for permitting the most controversial aspects of that closed-circuit process. The company received a total of fifteen permits from a range of national and local authorities.[27]

At the local and regional level, information exchange was marginally effective. Although communications between the Environment Protection Agency in Baia Mare and the Romanian Waters Authority were delayed, neighboring countries affected by the spill were informed in a timely fashion as required by the 1992 Convention on the Protection and Use of Transboundary Watercourses and International Lakes and by the Convention on Cooperation for the Protection and Sustainable Use of the Danube River.[28] According to these latter conventions, riparian parties shall inform each other without delay about any critical situation that may have transboundary impact or as soon as a sudden increase of hazardous substances is identified in the Danube River or in waters within its catchment area.[29] Even though communication on that international level was adequate, the local public often was not informed or was misinformed about the effects of the spill. Reports in the media often were not based on available facts and caused unnecessary distress among citizens.[30] Furthermore, there was insufficient information available even from the local authorities on the nature and consequences of the spill.[31] A positive result was that the accident did prompt the local public to exercise their rights and demand information about the safety of their own living space. This awakening of public awareness has been long in developing; water quality of the Tisza/Tisa and its tributaries has been affected by industrial and intensive agricultural activities for decades.

The NGOs involved in disseminating information about the spill were mostly international or Hungarian; organizations in Romania or Serbia do not appear to have played a significant role in educating the public locally or worldwide about the consequences of the spill. International organizations, such as the World Wildlife Fund, have succeeded in communicating the story of the spill worldwide, partly because of their already existing information networks around the globe and partly because they reacted to the news of the spill by quickly filling gaps in these networks. WWF also seized the opportunity not merely to report the story and the consequences of the spill, but also to place it in the context of a series of similar incidents in the recent past. By emphasizing that the Tisza/Tisa accident was neither unique nor unprecedented, NGOs highlighted the increasing need for proper and binding regulatory frameworks, international liability legislation, and an environmentally sound method of gold extraction. Thus, NGOs regarded the incident and the related communications not merely as a local disaster, but as a symptom of a worldwide problem with relevance to every country in which transnational companies are engaged in hazardous operations. Furthermore, the accident revealed that NGOs concerned about mining in the countries of CEE could cooperate across national boundaries in the future to monitor mining activities and to disseminate information to the public. Such cooperation could prove effective in influencing government decisions about future mining operations, since concerted NGO involvement from Romania and downstream countries like Hungary and Serbia could have a more powerful effect on governmental decision making than NGOs in one country acting alone.

Seeking policy solutions

On the international level, the United Nations Economic Commission for Europe (UNECE) has developed environmental legislation to prevent such international accidents. These include the Convention on the Transboundary Effects of Industrial Accidents, the Convention on the Protection and Use of Transboundary Watercourses and International Lakes, and the recently adopted Protocol on Water and Health.[32]

These legal instruments might have helped the affected countries—if the countries had adopted them. For example, the Convention on the Transboundary Effects of Industrial Accidents obliges its parties to develop cross-border contingency plans that immediately come into operation following an accident with transboundary effects. Of all the countries affected by the Tisza/Tisa spill, only Hungary had ratified this convention before the accident.[33] Ratification of the relevant UNECE conventions and protocols is an example of a mitigating strategy available at the international level.[34]

Given the magnitude of risk, policy responses need to focus not only on mitigation but also prevention. Industry has persistently opposed adoption of preventive strategies, while many in the NGO community call for restrictions or an outright ban on the use of cyanide in gold mining. Negative publicity and, presumably, political pressure have led to the formulation of a voluntary code of practice for such mining operations. Voluntary measures are by definition weak policy instruments, and as such do not carry guarantees that real improvements will occur, but they reflect industry's concern that completely ignoring the problem may lead to pressure for regulatory intervention. As a result, the International Council on Metals and the Environment (ICME) in partnership with UNEP developed a voluntary "International Cyanide Management Code for the Manufacture, Transport and Use of Cyanide in the Production of Gold."[35] The code involves monitoring and reporting on uptake and compliance; it is too early to assess its effectiveness.

Although there exist legislative models that could help prevent such accidents from happening in the future, resolution of *post facto* liability in such cases has been an unsolved problem for decades. Currently, there is no widely adopted, international legislation governing liability in such cases, even though Principle 21 of the Stockholm Declaration on the Human Environment states that nations are responsible for ensuring that activities within their jurisdiction or control do not cause environmental damage.[36] Furthermore, several international instruments proclaim the right of individuals to a healthy environment.[37]

In the 1986 Sandoz incident, Switzerland was not held responsible for breaching its international obligations to supervise emergency planning of Sandoz; claims for compensation were determined to be the sole responsibility of the company. By mid 1988, over 90 percent of the Sandoz claims had been processed and the victims compensated. However, since the compensation issues were solved in this particular case, no steps were taken to formulate more generic liability solutions that could be used in future situations.

In the case of Esmeralda Explorations of Australia, one of the principals in the Tisza/Tisa spill, one author has argued that Tisza/Tisa is a point in a pattern: "This disaster in Europe is only the latest in a long history of environmental and human rights problems caused by Australian mining companies around the world. It is clearly time for the Australian government to act to prevent further problems."[38] However, the structure of the legislation prevents the Australian government from being held legally responsible for the actions of Australian companies abroad. Nor has the Romanian government accepted responsibility, even though it issued permits for an operation that had insufficient risk management plans and proved to be unsafe. According to the World Wildlife Fund, Esmeralda Explorations Ltd. has attempted to go into administration or receivership in Romania following the Baia Mare spill in an apparent bid to avoid compensation claims.[39]

As this case history demonstrates, there is a pressing need for more strict environmental liability regimes to encourage companies to develop adequate risk management strategies and to ensure that states take appropriate measures to supervise potentially highly dangerous industrial activities.

The Individual Decision Maker

The commentary above traces the history of the Baia Mare incident and reflects on the policy implications. It demonstrates the widely understood fact that policies for international and, specifically, for transboundary actions are critical for societal well-being. However, such policies, when they are present, always are weighed, considered, and implemented (or not) by individual decision makers. This human context of environmental decision making is widely overlooked. That is, there is a culture that surrounds each person; that culture includes the individual's personal value set, history, and the institution in which he or she serves. That culture influences to some greater or lesser degree the one's freedom to take certain decisions, and one's willingness to exercise that freedom.

We argue that any attempt to understand and/or influence transboundary environmental policy must pay close attention to this more personal side of decision making. More specifically, the central argument of this paper is that discussions of sustainability, environmental training and policy development should incorporate explicit attention to this "man behind the curtain"—that is to say, to the role of personal values and decision making attributes that control the implementation of policies. This is especially important in cases of rapid social transition, such as in the CEE and the FSU, where decision makers who have been active in one cultural setting are being expected to function in another. The establishment of a new policy alone will not be sufficient to achieve a new view of risk and safety in decisions.

Notes

1. Throughout this paper, we use Tisza/Tisa to reflect the Hungarian and Romanian spellings.
2. P. Hardi, *Impediments on Environmental Policy-Making and Implementation in Central and Eastern Europe: Tabula Rasa vs. Legacy of the Past*, Policy Papers in International Affairs 40 (Berkeley: Institute of International Studies, University of California, 1992); Regional Environmental Centre (REC), *Strategic Environmental Issues In Central and Eastern Europe. Regional Report*, vol. 1 (Budapest: Regional Environmental Centre for Central and Eastern Europe, 1994).
3. BBC News, http://news.bbc.co.uk/1/hi/world/europe/790727.stm
4. Robert Moran, *More Cyanide Uncertainties: Lessons from the Baia Mare, Romania Spill—Water Quality and Politics*, MPC Issue Paper no. 3 (Washington, D.C.: Mineral Policy Center, 2001). http://www.zpok.hu/cyanide/baiamare/docs/mcu_final.pdf
5. Commission of the European Communities, *Communication from the Commission: Safe Operation of Mining Activities: A Follow-up to Recent Mining Accidents*. COM(2000) 664 final (Brussels: 23 October 2000), 3.
6. Ibid.
7. Robert Moran, *Cyanide in Mining: Some Observations on the Chemistry, Toxicity and Analysis of Mining-Related Waters*, www.mpi.org
8. N. H. Tarras-Wahlberga et al., "Environmental impacts and metal exposure of aquatic ecosystems in rivers contaminated by small scale gold mining: The Puyango River basin, southern Ecuador," *The Science of the Total Environment* 278 (2001): 239–61.
9. REC, *The Cyanide Spill at Baia Mare, Romania: Before, During and After* (Budapest: The Regional Environmental Center for Central and Eastern Europe, June 2000).
10. *Report of the International Task Force for Assessing the Baia Mare Accident* (December 2000). http://www.reliefweb.int/library/documents/eubaiamare.pdf
11. Commission of the European Communities, *Safe Operation of Mining Activities*, EC COM (2000) 664 final, p. 3. http://europa.eu.int/eur-lex/en/com/cnc/2000/com2000_0664en01.pdf
12. Commission of the European Communities, *Safe Operation of Mining Activities*.
13. Robert Moran, *Cyanide in Mining*.
14. Commission of the European Communities, *Safe Operation of Mining Activities*, 4.
15. World Wildlife Fund Press Release, *Another Dam Breaks—Still No New EU Law On Mining Waste*, 11 September 2000, http://www.panda.org/europe/freshwater/newsroom/newsroom17.html

16. United Nations Environment Programme (UNEP) / Office for the Co-ordination of Humanitarian Affairs (OCHA), *Cyanide Spill at Baia Mare, Romania: Assessment Mission Report* (Geneva: March 2000), 44.
17. UNEP/OCHA, *Cyanide Spill at Baia Mare*, 9.
18. William Drozdiak (Washington Post Foreign Service, Strasbourg, France), "Sewer of Europe" *The Washington Post*, 27 March 1996.
19. Ibid.
20. Ibid.
21. The Seveso II Directive, www.europa.eu.int
22. Ibid.
23. Drozdiak, "Sewer of Europe."
24. UNEP, *Rapid Environmental Assessment of the Tisza River Basin* (Geneva: 2004), http://www.grid.unep.ch/product/publication/download/tisza.pdf
25. *Report of the International Task Force for Assessing the Baia Mare Accident*, 9.
26. *Report of the International Task Force for Assessing the Baia Mare Accident*, 8.
27. UNEP/OCHA, *Cyanide Spill at Baia Mare*, 13.
28. UNEP/OCHA, *Cyanide Spill at Baia Mare*, 17.
29. Ibid.
30. *Report of the International Task Force for Assessing the Baia Mare Accident*, 21.
31. UNEP/OCHA, *Cyanide Spill at Baia Mare*, 43.
32. United Nations Economic Commission for Europe, *Cyanide Spill in Romania Wreaks Havoc on the Environment at Home and Abroad*, Press Release ECE/ENV/00/1 (Geneva, 15 February 2000).
33. Ibid.
34. *Report of the International Task Force for Assessing the Baia Mare Accident*, 30.
35. *International Cyanide Management Code for the Gold Mining Industry*, http://www.cyanidecode.org/
36. Karyn Keenan, *India, Switzerland and the United States: How Countries Avoid Liability after Disaster*, http://www.cowac.org/india.htm
37. Ibid.
38. Mineral Policy Institute, *"European catastrophe": Binding Regulation Needed To Repair Australia's Reputation* (14 February 2000), www.mpi.org
39. World Wildlife Fund, *WWF Challenges Mining Industry*, Press Release, www.panda.org

Chapter 9

AUSTRIA AND THE EU WATER FRAMEWORK DIRECTIVE

Wilhelm R. Vogel

Austria's accession to the European Union in 1995 had only minor consequences for water management. But in 2000, with the new EU Water Framework Directive (WFD), a new framework for Community action in the field of water policy was established, providing the basis for a restructuring of water management in all member states and setting new standards in terms of quality goals and water management approaches.[1]

For the Danube River basin, the development of one common river basin management plan for the whole area and involving member states, accession countries and others, if they agree to cooperate, is one of the most demanding aspects of this directive.

Austrian Law on Water Management before 1995

The basic regulations on Austrian water management are stipulated in the Water Act of 1959 and its amendments.[2] Wastewater emissions are limited according to the state of the art and may be formulated more strictly in individual licenses in case of small or sensitive receiving water bodies. These minimum requirements are described in detail (defining the state of the art) in sixty-six branch-specific ordinances on wastewater emission limits. Due to the application of the state-of-the-art principle, no detailed ambient water quality standards for rivers and lakes were considered to be necessary. Nevertheless, draft ordinances were developed based on the principle of ecological integrity stipulated in the Water Act. Formally, these ordinances on surface water quality goals never came into force, but they were generally used as guiding documents.

Groundwater is nearly the only source of drinking water in Austria (about 99 percent) and therefore is legally protected. According to the Water Act, groundwater should be of a quality allowing its use as drinking water without any further treatment. As a consequence, hardly any discharge from point sources into groundwater is licensed; however, leakages from old waste dumps, sewerage systems, and septic tanks are of significant influence. By far the most important diffuse source of groundwater contamination is agriculture, which contributes regularly to the relatively high concentrations of nitrates and pesticides. Financial incentives to promote environmentally sound management of farmland are the main instrument for improving the situation; applications include the reduction of fertilizer and pesticide use, encouragement of crop rotation, and development of organic farming.

The Impact of EU Regulation

Upon joining the EU in 1995, Austria had to implement practically the whole *acquis communautaire* in the field of environment as of the date of entrance. In the case of water management, this meant only minor modifications in the legislation and the obligations of reporting to the European Commission. The basic concepts remained untouched.

This changed significantly with the development of the Water Framework Directive.[3] This directive, which came into force in December 2000, is intended to provide a totally new basis for water management in the European Union. In Austria, the obligations of the Water Framework Directive were fulfilled by further amendments to the Water Act.

The EU WFD of 2000: The Basic Concept

The WFD is a product of negotiations between the European Parliament and the European Council (co-decision process) based on a text prepared by the European Commission. As a consequence, in some aspects, for instance the general philosophy of water management, the text is rather clear and ambitious. In other aspects, e.g. the limit values for groundwater contamination—where only a minimum consensus could be reached—the wording of the directive is not satisfying. The development of daughter directives is expected to add the necessary substance at a later stage.

The Goal of the WFD: "Good Status"

The WFD defines a goal and describes the process for reaching this goal. The basic goal is to reach "good status" in all water bodies.

For groundwater, this means "good quantitative status" and "good chemical status." Good quantitative status is reached where there is a more or less constant groundwater level and no adverse effects on dependent terrestrial or surface water ecosystems. Good chemical status is achieved if there are no adverse trends and there is compliance with other EU regulations that cover nitrates and pesticides.

For surface water, "good ecological" and "good chemical" status is required. The area of the European Union stretches from the far north to the Mediterranean, covering totally different climatic zones and necessitating that the system be applicable for a Scandinavian stream as well as for the estuary of a Mediterranean river. Such a system, based on the description of a reference status (situation nearly uninfluenced by man), is currently being developed in the member states for each surface water type.

Based on this system, good ecological status is achieved if the fauna and flora of a given water body do not differ very much from the natural state; the interpretation of the "natural state" has to be standardized by an intercalibration process among the member states. Good chemical status requires the achievement of chemical water quality that supports the biocoenosis of a reference situation.

The concept of a unique system of quality goals for the whole territory of the European Union and the associated reporting obligations will allow the European Commission to prepare maps showing the status of water bodies in the member states. It is rather likely that the mere existence of such a "map of Europe" will put significant political pressure on the governments of member states with poor water quality performance.

The Process Leading to "Good Status"

As in most European countries, water management in Austria used to be based on administrative geographical units, in this case the *Länder* or states. The WFD asks instead for the development of river management plans based on hydrological units—the river basins. This means cooperation between all the administrative units in one river basin. This cooperation is mandatory within one member state as well as between member states. In case of non–member states sharing a river basin with a member state, member states shall endeavor to establish appropriate coordination with relevant non–member states.

Consequences for Austrian Water Management

Austria has a long tradition in water management and, since 1992, runs a very elaborate water quality monitoring system[4] as a basis for "tailor-made" decision making.[5]

As for attaining the desired status, the concept of the "good ecological status" is the most demanding one. To implement it nineteen types of water bodies were described and delineated in Austria. The criteria for "good status" will be developed based on this classification.

The geographical basis for water management used to be the area of the provinces or states (see Figure 9.1). To comply with the river basin concept, cooperation mechanisms between the provinces and the central government were established to allow the preparation of appropriate river management plans without having to form new administrative bodies.

The territory of Austria has a share in three different river basins: those of the Danube, covering about 95 percent of the national territory, the Rhine, and the Elbe, the latter occupying only a few square kilometers (Figure 9.2). Austria thus will have to participate in the process of preparing the management plans for the Danube (eighteen states), the Rhine (nine states), and to a lesser extent owing to the small part of Austrian territory in this basin, the Elbe (four states).

The Management of the Danube River Basin

The development of a management plan for a river basin as a whole is most demanding in the case of the Danube River basin. This basin is shared by eighteen states (Figure 9.3), in which nearly the same number of different languages is spoken. Five of these states have only a small share in the Danube Basin (like Austria's share of the Elbe basin), leaving thirteen states to manage (and influence) significant parts of the Danube basin. To date, six of these states are member states of the European Union (Germany, Austria, and, since 2004, Czech Repub-

Figure 9.1 Administrative Units (Länder) in Austria

Source: Bundesministerium für Land- und Forstwirtschaft, Umwelt und Wasserwirtschaft (BMLFUW).

Figure 9.2 Basins and Sub-basins in Austria. One sub-basin belongs to the Rhine, one to the Elbe, and all others to the Danube.

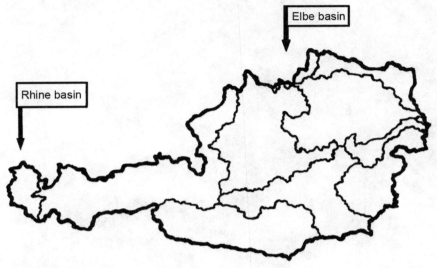

Source: BMLFUW.

lic, Slovakia, Hungary, and Slovenia), and negotiations have started with three others (Bulgaria, Romania, and Croatia).

For the implementation of the WFD, the Secretariat of the International Convention on the Protection of the River Danube, which is based in Vienna, is prepared to support the implementation and act as a facilitator by providing its administrative structure. For historical reasons the Austrian administration has a long tradition in cooperation with other states in the Danube basin and therefore is going to play an important role in supporting downstream countries.

The Time Scale

As a basis for the plan, various types of information have to be collected and managed: information on land use, pollution caused by point sources and diffuse sources, licenses for discharge and abstraction, drinking water supply and waste water treatment facilities, protected areas. Economic data are also needed to estimate the cost recovery of drinking water and wastewater management.

All of this has to be done according to a demanding time scale. River basin management plans have to be published in 2009. This process starts with the publication of a first concept in 2006, followed by a summary of the most important issues in 2007 and the preparation of a draft plan until 2008. A clearly

Figure 9.3 The Danube River Basin

Source: BMLFUW.

defined public participation process is linked to this development of the management plans, allowing six months for written responses at each stage.

In 2015, all water bodies of the European Union should reach the "good" status; a demanding goal—the clock is ticking!

Changes in Water Management Due to Implementation of EU Water Law (Lessons Learned)

It is too early to draw final conclusions concerning the effects of the implementation of EU law in the water sector, but first results are becoming apparent:

1. National and international cooperation is improving under the obligation to work on the scale of river basins.
2. Public access to information and public participation are slowly shifting the focus of water management.
3. The scientific basis of measures is improving due to the internationalization of national water management.
4. There is a strong incentive to improve electronic data management and information tools, in that the obligations of the WFD can hardly be met without state-of-the-art instruments.

5. Information on water quality is becoming more comparable throughout the territory of the EU, probably leading to more political relevance of the results.

Notes

1. General information concerning the European Union can be found at <http://www.europa.eu.int/>. The development of the water law is reflected in various relevant papers available at http://forum.europa.eu.int/Public/irc/env/wfd/library?1=/framework_directive
2. BGBl: 1959/215: Wasserrechtsgesetz 1959 (Austrian Water Act with several amendments) (Bundesgesetzblatt). Austrian legislation can be researched via http://www.ris.bka.gv.at/bundesrecht/
3. European Union: Water Framework Directive 60/2000/EC. EU legislation can be researched via http://europa.eu.int/eur-lex/
4. W. R. Vogel, "Water Monitoring in the Light of the EU Water Framework Directive—The Austrian Approach," in *Proceedings: International Conference on EU Water Management Framework Directive and Danubian Countries: Bratislava, 21–23 June 1999*, ed. Eva Pálmaiová (Bratislava: Stimul, 1999), 234–9; W. R. Vogel, "The Austrian Water Monitoring System and Its Integration into the Water Management Concept," in *Environmental Toxicology Assessment*, ed. Mervyn Richardson (London: Taylor and Francis, 1995), 227–46; C. Koreimann, J. Grath, G. Winkler, W. Nagy, and W. R. Vogel, "Groundwater Monitoring in Europe: European Topic Centre on Inland Waters," Topic Report 14 (Copenhagen: European Environmental Agency, 1996); W. R. Vogel, "The Water Quality Monitoring network in Austria—How to put Information to Work," *Proceedings of ECO-Informa, Neuherberg/Munich 6–9 October 1997* (Munich: Eco-Informa Press, 1997), 220–25.
5. W. R. Vogel, J. Grath, G. Winkler, and A. Chovanec, "The Austrian Water Monitoring System—Information for Different Levels of the Decision Making Processes," in *Proceedings: Monitoring Tailor Made*, vol. 2 (Nunspeet, 1996), 147–53.

Chapter 10

The Western Bug River
UNECE Pilot Project

James B. Dalton, Jr.

Introduction

The Western Bug River is an ideal river for studying the effects of a river's water quality management on the nature of a country's perceived water resource stress. This river's unique political and geographical position makes it a useful case for gaining additional insights into this connection. Politically, the river forms the border between Poland, a NATO and European Union (EU) member, and Ukraine and Belarus, non-NATO countries and currently not EU accession countries. These three Slavic countries, with different democratic histories and experiences with independence, are in transition from centralized planning and responsibility toward a democratic system.

The river is located in a geographic area that is susceptible to seasonal flow rates and dry climatic periods that stress water resources. The annual runoff from the Western Bug River, one measure of available surface water, is below the European annual average runoff. This paper illustrates that effective, proactive water quality management can help relieve stress caused by reduced water resources.

Approach

Essential to gaining an understanding of conditions as they pertain to this river are a review of documents concerning the study area and interviews with key water management individuals. Understanding water quality and quantity management of the Western Bug River will improve understanding of water resource stressors throughout the entire river basin. There are two primary sources of information, academic and governmental, whose bureaucratic boundaries separate their interests and activities. The government in general, and the Ministry of Environmental Protection, Natural Resources and Forestry (MOE) in particular, have been downsized in order to reduce a budget deficit and increase efficiency. In the academic arena, study concerning the Western Bug River has been concentrated at the Marie Curie-Skłodowska University in Lublin. Specifically, the Hydrology and Environmental Ecology Department takes a substantial leadership role in studies of the Western Bug because of its close proximity to the river. This department also has a long tradition of academic exchange with Lviv University, which also is located close to the river in Ukraine.

Why is this specific river important to Poland? First, one can trace the presence of a strong environmental movement in Poland from the early 1970s, but it was not until Poland's transition to democracy that observers noted an effective institutional response. The argument that it was this transition that added credibility to the environmental movement can be traced to several essential documents from that period.[1] As early as March 1989, during the Round Table talks between the coalition-communist government and the opposition-Solidarity organization, the environment was a major concern that warranted its own subunit on ecology. The Round Table report recognized transboundary sources of pollution and directed that "efforts be made to ensure cooperation and co-action of ecological organizations of neighboring countries."[2] Following these negotiations, the Constitution of Poland was ratified; it included Article 5, which declared: "The Republic of Poland protects its independence and inviolability of its territory, ensures human and citizen liberties and rights as well as its citizens' safety, stands on guard of its national heritage and ensures environmental protection in accordance with the sustainable development principle."[3] Two years later, the National Environmental Policy of Poland stated that the Western Bug River was one of the transnational rivers whose water pollution reaches Polish territory, and that areas close to Poland's borders with poor environmental conditions pose a danger to Poland. The policy further stated that diplomatic and environmental authorities should intensify services aimed at protecting Polish interests from these cross-border dangers.[4]

A second reason the Western Bug warrants special attention is that despite

> a dense hydrographical network, very shallow occurrence of groundwater, and the existence of 68 lakes, bogs and swamps of all kinds that create the impression of water richness of this region, it is an area of water scarceness. The above features are not due to an ample supply of water but to the occurrence of shallow impermeable layers that make water circulation difficult.[5]

The Basin's Territory

The Western Bug River is situated in Central Europe, with its headwaters originating in the northwesternmost corner of the Podolian Plateau, specifically in the Miodobory Hills near Lviv, Ukraine (Figure 10.1). It is distinct from the Bug River

Figure 10.1 Relief Map of Bug River Basin

Source: University of Minnesota Cartography Lab.

that originates south of Kiev in Ukraine and flows generally south to the Black Sea. From its headwaters, the Western Bug runs unobstructed by dams for most of its 772 km. The river flows from its origin to the Ukrainian-Polish border, then forms the border between Poland and Ukraine for 205 km before becoming the border between Poland and Belarus for 158 km. At its lower end, the river flows into the man-made Zegrzyńskie Lake and finally into the Vistula (Wisła) River.

The basin covers a total area of 39,420 km² with 23.4 percent in Belarus, 27.4 percent in Ukraine, and 49.2 percent in Poland. Fed by twenty-two major tributaries, the Western Bug joins the Narew River to flow into the Vistula River northwest of Warsaw.[6] While here the emphasis is on Poland and Ukraine's relations concerning the Western Bug River, there is also considerable discussion concerning Belarus in order to provide the overall context. Over three million people live within the Western Bug's catchment. Almost two million live in Ukraine, a half million in Belarus, and over a million in Poland. Additionally, the city of Warsaw, with a population of 2.9 million, relies on Zegrzyńskie Lake as one of its main sources of drinking water.[7]

This region's climatic conditions produce a pronounced dry period in autumn, reducing water availability throughout the entire river basin. In Poland, a country with a 1,500 m^3 per capita water supply annually, water is relatively scarce. Compared with other European countries, Poland has the seventh lowest water availability while it has the sixth largest population. Poland's water is also spatially and temporally spread unevenly throughout its territory, making all of Poland extremely susceptible to dry periods.[8]

To varying degrees, Ukraine, Poland, and Belarus share certain sources of pollution. Ukraine's major pollution sources are industry, agriculture, mining, petroleum, and wastewater from municipalities. Poland cites agricultural, mining, and wastewater discharges as its major contributions to the river's pollution. The discharges from mining operations are extremely saline and highly corrosive, which particularly affects Poland and Ukraine.[9] Belarus contributes pollution from industry, agriculture, and municipal wastewater to the Western Bug River.[10] Additionally, as downstream countries, Poland and Belarus list inflows from Ukraine as a major negative impact on their water quality.

Potential Basin Problems

The previous paragraphs highlight some of the general problems facing the river system. Taken together, a general picture emerges of a river basin containing a slow meandering stream in which the various sources of pollution have accumulated to produce conditions that led Polish water resource personnel to designate this river with their lowest classification. What is important in this discussion of the study site is to realize that flow rates (quantity), either high or low (see Figure 10.2), affect water quality. Specifically, a low flow rate increases concentration of pollutants, whereas a high flow rate including increased runoff from agricultural land increases pollutant loads and damages the stream bed of this normally slow moving river with rushing water. In addition to quality and quantity considerations, these environmental conditions also pose potential problems for cooperation.

Political Overview

Three different countries manage the Western Bug basin, each at a different stage in its development of effective management systems and procedures for its por-

Figure 10.2 Monthly Flow Characteristics for the Western Bug River at Wyszkowie from 1951 to 1990

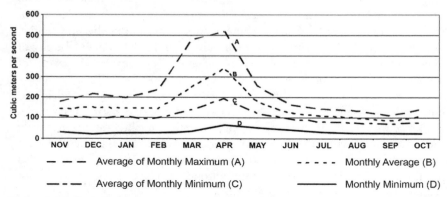

Source: (Michalczyk forthcoming).

tion of the river basin. Each country described its method of management in a recent interim report to the Task Force on Monitoring and Assessment under the United Nations/Economic Commission for Europe (UNECE) Water Convention Pilot Project Programme Transboundary Rivers.[11]

What follows is a short description of the state of water management within each country sharing the basin. Belarus is illustrative of the different governmental scales involved. Water is monitored at the national, regional and local level in Belarus, with a total of fifty one water-sampling points.[12] The national points are run by the State Hydrometeorology Committee, with eleven points for measuring water quantity and six for water quality. The Brest Regional Committee for Natural Resources and Environmental Protection operates the six regional points; four of these points are joint observation points with Poland. All six regional points are water quality measuring points. Water quality is measured using both "hydro-chemical and hydrobiological parameters" with a total of sixty-four measurements taken for reporting. However, the water pollution index utilized to describe the state of the water consists of only six parameters of water chemistry: dissolved oxygen, BOD_5, ammonia, nitrites, petroleum products, and phenols.[13] Based on this index, rivers can be classified from Class 1 (very clean) to Class 7 (extremely polluted). Belarus ranks the Western Bug as a Class 3 river (moderately polluted).[14]

Ukraine manages its surface water, including the Western Bug River, through the combined efforts of five ministry-level organizations. While currently in a transitional stage, the system nevertheless functions under the coordination of the Ministry of Environmental Protection and Nuclear Safety, which has overall responsibility for monitoring and assessment. However, the actual implementation of policy is carried out by different organizations at national and regional levels.[15]

The 1998 UNECE State of Environment Report for Ukraine found that the organizations with environmental protection responsibilities were not efficient, and that overall operations could be improved with more centralized, streamlined organization. The report determined that if Ukraine's Resolution No. 391 of 1998 were fully implemented, and if the Ministry of Environmental Protection and Nuclear Safety were responsible for coordinating all monitoring activities, then operations would be improved.[16]

Unlike Ukraine, which has been independent only since 1990, Poland has been measuring its river pollution for the last thirty-five years and publishing the results in numerous river atlases. Poland's monitoring system is run by the Chief Inspector of Environmental Protection, who reports to the Minister of Environmental Protection. The monitoring system is well established, with testing frequency set according to the class of indicator. The sampling and analysis are conducted by the State Inspectorate of Environmental Protection (SIEP) and will be discussed in greater detail later. The two areas that are not well developed in Poland are the early warning monitoring system and the monitoring of pesticides.[17]

Regional Management

The foregoing physical, political, and geographical circumstances place Poland in a position to act as a leader in the development of a system of water quality and quantity management that will increase Poland's environmental quality, stability, and security. In order to accomplish this goal, Poland has individuals, both within the government and in academic institutions, working on the Western Bug River at the regional and national levels. These individuals are the key to Poland's environmental future, as they work through the maze of obstacles that present themselves in the course of managing a transnational river. The impediments to stable, cooperative management of the river are varied and evolving. Managers are challenged throughout the iterative, continuous process.

The Polish regional officials managing the Western Bug River are located in the Lublin Voivodeship (province). They operate in two independent departments: the Regional Water Management Board-Lublin Division and the Voivodeship SIEP. The Regional Water Management Board is responsible for detailed water management planning, water conservation, erosion protection, exploitation, and inventory of water users and their discharges. The board's assessment of the water balance in the basin is still in progress. It will use a geographic information system (GIS) to track the inflows and outflows of the river and help to formulate its limits and capacities.

A recent reorganization within the Ministry of Environmental Protection, Natural Resources and Forestry, has divided responsibility for and management of the Western Bug River basin among three different sub-regional water management boards. This study considers just the Lublin Headquarters Office, which handles the stretch of the river that forms Poland's border with Ukraine. Members of this subregional water management board are very active in managing the Western Bug River and have many years' experience working in this region. They report to the Regional Water Management Board in Warsaw.

The other major regional office is the Voivodeship office of the State Inspectorate of Environmental Protection (SIEP). Under the recent reorganization, this office now reports to and receives funding from the Lublin Voivodeship government. This office is responsible for monitoring, assessing, classifying, and reporting on the water quality of all surface waters in the voivodeship. It operates three labs to conduct analysis of samples taken from surface waters at regular intervals. There are five established monitoring sites on the Western Bug where it forms the border with Ukraine, and five monitoring sites where it forms the border between Poland and Belarus.

National Stewardship

The Western Bug's management at the national level is conducted primarily by two departments within the Ministry of Environmental Protection, Natural Resources and Forestry. Within this ministry the Department of Water Resources is responsible for national water quality planning and for cooperation with neighboring countries concerning shared water resources. The current acting deputy director of the Department of Water Resources is also the Polish Commissioner of the Polish-Ukrainian Commission on the Western Bug River. The other national office involved, the Monitoring Department within the Chief Inspectorate for Environmental Protection, is responsible for national monitoring policy and collection of data from all sixteen voivodeship SIEPs. Concerning the Western Bug River, these national agencies collect data from the Lublin and Warsaw Voivodeship inspectorate offices.

The Transnational Level

The European Union (EU) has a significant impact as a transnational organization whose water quality standards must be met by member and accession nations. As stated by the Dobris assessment:

> Water management should be linked to catchments, and be carried out with the participation of a well informed public:
> - Strengthening the provision of education and information.
> - Further development of structures and procedures for solving conflicts between competing water users, at national as well as European level.[18]

Poland, as an accession nation, is responsible for developing legislation that will bring its water law and water management regulations and practices into conformity with EU standards and directives. The executive agency for the EU here is the European Environmental Agency (EEA). The two key pieces of legislation developed by the EU are the Water Framework Directive (WFD) and the Waste Water Directive. Together, these documents spell out the requirements for action by member and accession nations and are major policy responses to water quality stress.[19] These directives are designed to align with the UN's Agenda 21.

Additionally, the EEA has begun a river monitoring program, called EUROWATERNET,[20] which when complete will allow for a Europe-wide river water quality assessment. This program is one outcome of the WFD, which requires the integration of "water resource management with the protection of the natural ecological state and functioning of the environment; water quality and water quantity management; surface water management (including coastal waters) with ground waters management; measures such as emission controls, with environmental objectives."[21]

An addendum to the WFD requires the monitoring and assessment of the ecological and chemical status of surface waters and the quantitative and chemical status of groundwater.[22] These requirements place additional responsibility on accession nations to conform to the EU rules, and carry with them financial as well as scientific obligations.

The UNECE in collaboration with the concerned countries has had an impact on the Western Bug River and seven others through its pilot project on Monitoring and Assessment of Transboundary Rivers. This program is designed to test monitoring and assessment practices that will be transferred to other rivers with similar conditions. The program's main goal is to "demonstrate the Guidelines on Water-quality Monitoring and Assessment of Transboundary Rivers as drafted by the Task Force on Monitoring and Assessment under the Convention on Protection and Use of Transboundary Watercourses and International Lakes."[23]

NATO's role in promoting water quality in Poland specifically is minor, but it plays a more general role in defining the terms of environmental security. A NATO pilot study entitled "Environment & Security" by NATO's Committee on the Challenges of Modern Society (CCMS) sets out NATO's concept of environmental security and the role water plays within that context. In particular, the study calls for proactive measures that will reduce the likelihood of tension or conflict arising from an environmental stressor. It reports that one factor of environmental stress is the "scarcity of a renewable natural resource which involves the reduction or perceived reduction in total quantity or available quantity of natural resources such as arable land, fresh water, forest or fish stock."[24]

Polish Water Quality: How Is It Determined and By Whom?

Water quality in Poland is assessed using a set of fifty-two biological, chemical, and physical indicators. The selected indicators, developed over several decades, form the basis for classification of surface waters. They date back to a 1972 law for water quality, which was subsequently updated in 1991. The 1991 law, Dziennik Ustaw Nr. 116, set the new standards for water quality classification. Water quality in a river segment is classified at one of four different levels: I (best), II (next best), III (least desirable), and below III (extremely poor). Every one of the fifty-two indicators has equal weight, so that each river segment is classified according to the worst indicator classification. For example, if a river section has one indicator that is Class III but all other indicators are Class II, that section is classified as Class III. The points along rivers where indicators are measured form part of three river monitoring networks controlled at the national and regional levels.

At the national level, there is a network consisting of three categories of measuring points for rivers. The largest group has 361 basic points located on the most significant Polish rivers. The second category consists of the border river sampling points. The third national category covers twenty special points called geodetic points located on rivers that discharge into the Baltic Sea. Two organizations take the actual samplings at these national points. Personnel from regional offices of the SIEP handle the first and second categories, and the third category is the province of the Meteorological and Hydrological Institute (IMGW) personnel.

At the regional level, additional points are added to the national network based on the needs of the offices of the regional SIEPs. Each office has the discretion to place additional monitoring points on any of the surface waters within its voivodeship. At the local level, users monitor their discharges and report those measurements to the regional SIEP offices and the Regional Water Management Board. The frequency of sampling accords with guidelines established in enclosures Four and Five to the 1991 water quality law. These provisions categorize the fifty-two indicators into three groups and then specify the minimum frequency acceptable for each group. Regional administrators have the authority to increase the frequency of measurements if the SIEP concludes that the situation warrants additional measurement. Even though it is not legally binding on voivodeships to use these standards, in practice they are used nationally for classification. Water quality managers anticipate that these national standards will become the legal standard for compliance with the EU accession requirements mentioned earlier.[25]

Local-scale evaluation of water quality currently does not play a part in the creation of standards nor the sampling and assessment of water quality in Poland. One water quality official stated that the long-range goal is to set up local water quality offices but did not see that happening in the short term due to financial constraints.

The regional scale is the most active framework for the actual sampling, monitoring, and assessment of the Western Bug River. It is at this level that samples are taken, analyzed, and compiled to obtain a snapshot of the river in accordance with the prescribed sampling frequency. At the end of each year, the data are aggregated to create an assessment of the river for the year. This annual assessment is used for internal evaluation of sampling points and frequency, and as a report to national water authorities and the general public. At the regional level, the report covers all the surface waters of the voivodeship. A series of maps is created to summarize the conditions of the river. All major surface waters in the voivodeship are classified according to the methods described earlier. The report highlights the Western Bug River—specifically, at the international border points—because of its importance to the voivodeship. The maps and charts in the report are vital information sources for regional water quality management and inform the national water quality managers about the Western Bug's water quality.

National/Transnational

At the national scale, water managers use the information provided by the regional offices of both SIEP and Regional Water Management Board authorities to determine national water quality. Their results are used by decision makers and academics and are also released to the public. At the national level, the bulk of the Western Bug River is classified as below Class III.[26] This classification results from a combination of data from SIEP offices in two different voivodeship, and the Regional Water Management Board. In this way, regional and national officials have united in an effort to manage the water quality of this transnational river. The outcomes of these efforts can be helpful in evaluating the degree of cooperation that is actually taking place.

Interactions between Polish and Ukrainian officials concerning the Western Bug River take place on multiple scales, among both regional and national water managers. The interaction on the national scale is guided by the Polish-Ukrainian Bilateral Commission for the Problem of Border Water Matters, which is the chief mechanism for interaction on a national scale. Formed upon the basis of a bilateral agreement signed in 1996 and ratified in 1999, the commission is charged with developing and implementing synchronized management of the Western Bug River's upper reaches. The commission is composed of joint commissioners with national staffs and five working groups. These working groups, each charged with a specific area of concern, provide the mechanism for regional-scale interaction and will be discussed further in the next section.

The first meeting of the commission took place in early 2000 and worked to establish the mandates and procedures for the commission.[27] The Polish commissioner has been authorized to speak for the Polish government; as of August 2000 it was proposed that this official also be authorized to make decisions concerning the river on behalf of Poland. The Ukrainian Commissioner, based in Kiev, reports to Ukraine's Minister of Environmental Protection and Nuclear Safety for approval of all actions.[28]

The other major form of interaction between Poland and Ukraine in environmental matters, as mentioned earlier, is the Task Force on Monitoring and Assessment under the UNECE Water Convention Pilot Project Programme Transboundary Rivers program that is being implemented at the national level by the three riparian nations with sponsorship of the UNECE. The program's main goals are to demonstrate effective and efficient monitoring and assessment of water quality in transboundary water basins, and to establish specific principles in monitoring and assessment on a number of rivers in Central and Eastern Europe.[29] These specific goals are supportive of additional, wider UNECE program goals "to initiate and/or improve bilateral and multilateral co-operation, leading to institutional strengthening and capacity building under the convention; to support the approximation [alignment] to the EU environmental legislation in CEEC-countries."[30] Even though this program was started earlier than the founding of the bilateral commission, it has become an important part of the commission's work and is seen by Polish members as a significant means of cooperation between Poland and Ukraine. People at the national and regional level are working to fulfill this program's goals.

The working groups of the joint Western Bug River commission are the regional means of interaction and cooperation. Each group is composed of experts from both nations with the same level of responsibility and, in theory, the same level of authority. Even though each working group has Polish and Ukrainian members from regional offices, it also comprises national representatives from the appropriate governmental offices. The groups are named after their areas of responsibility; a shorthand nomenclature of two letters to identify each group was agreed upon in the official accord. The groups and their respective spheres of responsibility are Group PL, for strategic planning along the border waters; Group OW, protection of the border waters from pollution; Group OP, for flood protection, water control, and irrigation; Group NZ, for the planning and preparing for environmental emergencies and recovery; and Group HH, for hydrometeorology and hydrogeology of the river.

A review of all the groups' combined actions indicates that, as a combination of activities, the process itself helps to facilitate decreased water resource stress as defined by the Polish Ministry of Environmental Protection, Natural Resources and Forestry. However, this process could be delayed by the following potential obstacles: (1) inability of the working groups to meet and reach some level of cooperation, (2) insufficient funds to conduct the work necessary to achieve their goals, (3) insufficient administrative flexibility on the part of the Ukrainian bureaucracy, or (4) a lack of data sharing that could result from any combination of the first three potential obstacles. These possible roadblocks could threaten future cooperation between these countries and thereby diminish the effectiveness and benefits of environmental stress reduction.

Effectiveness

The effectiveness of the river commission and the pilot monitoring program for the Western Bug River can be evaluated first in terms of the amount and frequency of meetings conducted, and the effectiveness of communication. Second, the amount of data and timeliness of data sharing can be used to indicate the degree of cooperation that is taking place. A third measure of effectiveness is the actions taken as a result of the programs. The last three years have shown an increase in the number of interactions between Poland and Ukraine. After a slow start, meetings are taking place under the auspices of the two programs discussed earlier. Specifically, the commission's working groups meet regularly.

The pilot program for Monitoring and Assessment of Transboundary Rivers is run in conjunction with the commission's work, which further strengthens the cooperation and coordination between these two countries at the regional and national level. The data sharing that has occurred as a result of these activities is still confined within a slow and heavily layered system, but there is tremendous potential for improvements in the future. Current accounts indicate a deficit in timely data sharing from Ukraine to Poland, especially from the upper reaches of the river.

The activities associated with these programs will increase as the agreements and the mandated objectives become more fully implemented. Program activities will need continued monitoring, and their regularization will be a sign of progress in the joint management of this transnational river. The cooperation developed and the knowledge gained through these programs will work to reduce perceived water resource stress. At the national level the perception of water resource stress for the Western Bug River is already low due to the relative stability of the river. The regional concern some express over unknown water pollutants will certainly be allayed if these programs continue. Regional officials also recognize that, in the long run, Poland needs to incorporate the local level in its water management as stated in the Polish National Environmental Policy (1991, 2001).[31] A review of the total impact of the two major programs on the perceived water resource stress in Poland demonstrates that these programs can and will reduce the perception of environmental stress.

Conclusion

The preceding findings indicate that the agreements between Poland and Ukraine, because of their proactive and preventive nature, show promise for increasing the stability of this region and create the potential for an effective partnership in managing this transnational river, once fully implemented. A summary of the two Polish-Ukrainian agreements (Table 10.1) concerning the Western Bug River helps to illustrate this promise. Aspects of the agreements are displayed in with columns listing their purpose, stage of implementation, strengths and weaknesses, relevant comments, and current conditions' potential impact on water resources stress. The programs, listed by row, are examined with respect to different levels of organizational scale. On all but one organizational scale, the current conditions favor a decrease in water resource stress.

The countries have started to forge a solid basis for cooperation. The Polish assessment is that the current stability of the Western Bug River (i.e., the fact that it is not getting any worse) will buy time for both parties, time to work through the challenges of full implementation of the agreements.[32]

As recent history has shown, these agreements are susceptible to a wide variety of potential dangers. Most recently, mild (1995) and severe droughts (1990, 1994) caused low flow events that increased pollution concentrations. Addition-

162 | James B. Dalton, Jr.

Table 10.1 Summary of Major Water Management Programs for the Western Bug River

Program:	Purpose:	Stage:	Strengths	Risks/Weaknesses	Comments	*Impact
UNECE Monitoring and Assessment Pilot Program	Demonstrate the implementation of the UNECE guidelines for monitoring and assessment	Each river basin at a different stage	Vehicle for cooperation and coordination	Individual countries must find additional funding themselves if own budgets will not suffice.	Conducted on eight rivers over three phases. Guidelines to be revised upon completion.	decrease
Transnational:		Phase I completed Phase II started			Program accepted by all countries involved; began in 1997	decrease
National: Poland		Phase II	Nested into commission on Border Waters	Large amount of financing from own resources—competes with other environmental programs		decrease
National: Ukraine		Phase II	Receiving outside assistance in funding, collection and analysis of data	Slow implementation, data collection analysis and dissemination by another country		decrease
Regional: Lublin		Phase II	Using methods that have been developed and implemented over a long period of time	Finances and staff are thin	Encourages data sharing	decrease
Regional: Lviv		Phase II	Receiving help from outside sources	Adds another layer of coordination	Encourages data sharing	decrease

Organizational Scale

UNECE Pilot Project on the Western Bug | 163

Program:	Purpose:	Stage:	Strengths	Risks/Weaknesses	Comments	*Impact
Bilateral Agreement on Border Waters (Poland-Ukraine)	Cooperation in developing water management plans	Signed and ratified	Multifaceted, covers broad areas of water quality and quantity management.	Financial capability to implement is weak.		
Transnational:		Implementation				decrease
National:		Commission established and meeting regularly	Functional area orientation; incorporates UNECE Pilot program; basin organization; established working groups, decentralized	Ukrainian bureaucratic layering may slow progress; national environmental goals/requirements compete financially		decrease
Regional:		Working groups meeting	Combines experts and administrators; has developed groups mandates for national approval	Staff undermanned		decrease
Local:		None	n/a	Absent—risks losing local support without local participation		increase

*decrease = potential to decrease stress
increase = potential to increase stress

Sources: Developed from UNECE Inception Report Number 1 (1998) and Polish-Ukrainian Bilateral Agreement on Border Waters (1999).

ally, the floods of 1998 and 2001 are prime examples of the disruptive and damaging consequences of high water flows. Human-induced environmental disasters caused by accidental spills of various types are always a potential danger to the river basin, considering the industrial development in the upper reaches of the river in the vicinity of Lviv, Ukraine. These dangers can be minimized as these programs are fully implemented and planning for contingences is completed.

The detailed and consistent water management program in use in Poland illustrates Polish commitment to improving the country's water quality, and the Polish water classification system reflects high standards for water quality. Poland selects and publishes the Polish water standards for its surface waters. These published documents help inform the general public and contribute to perceptions of water quality in Poland. Further, this information demonstrates the Polish drive to meet EC-WFD goals.

Lastly, the work accomplished on this river basin provides a further model for other countries to follow as they seek to comply with the numerous water quality and quantity directives.[33] The countries that share the river basin have taken concrete steps to improve it even though only one will be an EU country in the near future. UNECE Pilot Project Report Number 2 clearly articulates these issues and concerns, providing a framework for others to consider in managing their own transnational river basins. The UNECE pilot project continues to provide a vehicle for cooperation and funding assistance for these countries.

Notes

1. Zbigniew. Bochniarz and R. S. Bolan, *Institutional Design for Financing Sustainable Development: Lessons Learned from Poland* (Minneapolis: University of Minnesota, 2000), 1–23.
2. *Report of the Round Table Subunit on Ecology* (Warsaw: Government and Solidarity Opposition, 1989), 1–29, here p. 18.
3. Data on the state of the environmental and economic situation in Poland have been taken from materials published by the Environmental Protection Inspectorate and the Main Statistical Office. I found this material at http://www.mos.gov.pl/mos/publikac/Raporty_opracowania/guidelines /index.html
4. Ministry of Environmental Protection, *National Environmental Policy of Poland* (Warsaw: Ministry of Environmental Protection, Natural Resources, and Forestry, 1991), 1–24, here pp. 22–3. The National Defense Strategy of the Republic of Poland clearly states that environmental protection is one of four national interests. Specifically, this document provided the impetus for the development of a department of Environmental Security of the Infrastructure within the National Defense Ministry. See also "The Environment in Poland," report by Colonel Marszalik, Chief, Environmental Security of the Infrastructure.
5. Z. Michalczyk, "Dangers for the Polesie Lubelskie Waters," in *The Regional Ecological Problems, Lviv, Ukraine* (Lublin: Maria Curie-Sklodowska University Press, 1996), 1, 83.
6. The main tributaries flowing from Ukrainian territory are the Dumni, Gapa, Luga, Poltva, Rata, Solokiya, and Studianka; from Belarus territory the Kopalovka, Pulva, Lesnaya Pravaya, Lesnaya, Muhavetz, Ryta, and Maloryta; from Polish territory the Sołokija, Huczwa, Uherka, Krzna, Kaminanka, Nurzec, Brok, and Liwiec.

7. V. Bilokon, B. Fornal, and A. Samusenko, *Bug: Report No. 1: Inception Report* (Zamosz: Task Force on Monitoring and Assessment under the UN/ECE Water Convention Pilot Program Transboundary Rivers, 1998), 8, 9, 43.
8. Z. Kaczmarek, J. J. Napiorkowski, and D. Jurak, *Impact of Climate Change on Water Resources in Poland,* Institute of Geophysics, Polish Academy of Sciences no. 295, Series E-1 (Warsaw: The Institute of Geophysics, 1997), 3.
9. European Environmental Agency, *Environment in the European Union at the turn of the century* (Brussels: European Environmental Agency, 1995), 3.
10. Belarus is faced with additional challenges in regard to its water management, simultaneously trying to resolve its own political course for the future as well as to maintain current bilateral agreements. Compounding the situation is the increase in agricultural activity in the vicinity of the Western Bug that has resulted from the 23 percent loss of agricultural land due to the Chernobyl disaster; Bilokon, Fornal, and Samusenko, *Bug: Report No. 1.*
11. Ibid.
12. M. Landsberg, *Bug: Report No. 2: Identification and Review of Water Management Issues* (Zamosz: UNECE Water Convention, 2002), 57.
13. Bilokon, Fornal, and Samusenko, *Bug: Report No. 1,* 11, 12.
14. Ibid., 32.
15. Ibid., 13.
16. United Nations Economic Commission for Europe, *EPR Recommendations made to Ukraine by the UN/ECE Committee on Environmental Policy* (New York and Geneva: United Nations Economic Commission for Europe, 1999), 47.
17. Bilokon, Fornal, and Samusenko, *Bug: Report No. 1,* 12, 42–7; I. O. Srodowiska, *State of the Environment for 1998,* Lublin Voivodship (Lublin: Inspectorate for Environmental Protection, 1999), 1–256.
18. European Environmental Agency, *Environment in the European Union at the turn of the century* (Brussels: European Environmental Agency, 1995); David Stanners and Philippe Bourddeau, eds., *Europe's Environment: The Dobris Assessment* (Copenhagen: European Environment Agency, 1995).
19. K. Lanz and S. Scheur, *EEB Handbook on EU Water Policy under the Water Framework Directive* (Brussels: European Environmental Bureau, 2001), 169–80.
20. EUROWATERNET is the European Environment Agency's Monitoring and Information Network for Inland Water Resources.
21. S. Nixon, J. Grath, and J. Bøgestrand, *EUROWATERNET, The European Environment Agency's Monitoring and Information Network for Inland Water Resources* (Copenhagen: European Environment Agency, 1998), 9.
22. Ibid.
23. Bilokon, Fornal, and Samusenko, *Bug: Report No. 1,* 2.
24. K. M. Lietzman and G. D. Vest, *Environment and Security in an International Context* (Bonn and Washington, D.C.: North Atlantic Treaty Organization, 1999), 174, 967.
25. In terms of monitoring requirements, a comparison of Polish and EU water management standards shows that Poland will be in compliance with EU standards once the current Polish standards are made legally binding. In fact, the Polish standards are more difficult to achieve than many of the EU standards. A problem identified by Polish water quality managers is the lack of consensus on a clear EU water quality standard and on consequences to EU members in case of noncompliance. A recent news report clearly identified the ways in which Polish industries will be hard pressed to meet the EU standards for emissions. The EU is providing €180 million for Polish businesses to help with achieving EU standards; see Izabella Kamińska, "EU Environmental law too much too soon for Poland," *Warsaw Business Journal,* 29 July 2002. Article published on the web by New World Publishing Kft and New World Publishing, Inc. 2002, at http://courses.wcupa.edu/rbove/eco343/023Compecon/Centeur/Poland/020729EU1.txt

26. Z. Kamienski, *The State of the Environment in Poland* (Warsaw: The State Inspectorate for Environmental Protection, 1998), 175.
27. P. Rutkiewicz and W. A. Sergijowicz, "Agreement between the governments of The Polish Republic and Ukraine on Cooperation in the Field of Water Management on Border Waters" (Kiev, 1996), 7. This is the formal government agreement between Poland and Ukraine, signed in Kiev.
28. Ibid.
29. United Nations Economic Commission for Europe, *Protection of Transboundary Waters, Guidance for Policy- and Decision-making* (New York and Geneva: United Nations Economic Commission for Europe, 1996), 36; Bilokon, Fornal, and Samusenko, *Bug: Report No. 1*, 4.
30. Bilokon, Fornal, and Samusenko, *Bug: Report No. 1*, 4.
31. Ministry of Environmental Protection, *National Environmental Policy of Poland* (Warsaw: Ministry of Environmental Protection, Natural Resources, and Forestry, 1991, 2001).
32. S. Zecchini, *Environmental Performance Reviews—Poland* (Paris: Centre for Co-Operation with the Economies in Transition, Organization for Economic Co-operation and Development, 1995), 185; Kamienski, *The State of the Environment in Poland*.
33. There are a total of fourteen directives for Poland to implement. The most notable are the "Bathing Water Directive," "New Drinking Water Directive," "Nitrate Directive," and "Urban Waste Water Directive."

Chapter 11

WASTEWATER TREATMENT IN THE POSTCOMMUNIST DANUBE RIVER BASIN

Igor Bodík

Introduction

In the aftermath of historical developments during World War II, the majority of the European communist countries were formed in the Danube River basin. The environment of the Danube basin has been under significant pollution stress for several years. Industrial wastes were, and still are, often disposed of or emitted without due consideration to the environment. Many of the municipal wastewater discharges that often contain a high industrial portion continue to flow without treatment to reduce polluting loads.

After more than forty years of economic mismanagement and environmental neglect, these countries have started to correct the effects of the previous rulers' policy in this field. As far as environment and water pollution are concerned, the legacy from the past regime is serious. It is characterized by a high level of water contamination, and the coexistence of problems caused by traditional pollutants as well as point and nonpoint sources. The additional difficulties caused by the past contamination of soil, sediment, and groundwater present the issue of a costly and slow rehabilitation.

In spite of the features mentioned above, the water pollution problems of the postcommunist Danube countries should not be considered unique in a technical sense. Similar situations existed in industrialized regions of the West (e.g., the Ruhr and Rhine rivers in Germany) about thirty years ago, and there is evidence that tools and technologies are available for the cleanup. The uniqueness stems from the coincidence of the need to handle these weighty environmental issues along with the very specific political, economic, and social conditions of postcommunist development.

The main objective of this study is to analyze the recent status of water management in the Danubian countries with a focus on sewage and municipal wastewater treatment in postcommunist countries in this region. Some detailed information about wastewater collection and treatment will be presented for the Slovak Republic.

Basic Characteristics of the Danube River Basin

Altogether, there are seventeen European countries occupying part of the Danube River basin (DRB)(see map, Figure 11.1). Because the part of the territory of the DRB (1.8 percent) that belongs to Italy, Switzerland, Poland, and Albania is insignificantly small, these countries will not be included in this study. Some basic geographical and economical indicators for the thirteen Danube countries are compiled in Table 11.1.

Table 11.1 indicates that significant portions of the territory and population of the DRB are formed by former "communist bloc" countries. The socioeconomic differences among the Danube countries are evident from the Gross Domestic Product (GDP) values. These discrepancies have to be considered as a substantial hindrance to implementing a balanced and optimally cost-effective water pollution reduction program for the Danube River system.

Figure 11.1 Location of Danube River Basin Countries

Table 11.1 Basic Geographical and Economic Parameters of the Danube Countries

Country		Present population	Present population in the DRB		Country territory	Country territory in the DRB		GDP in 1997	Inflation rates in 1997
		Mil.	Mil.	%	1,000 km²	1,000 km²	%	USD/cap.	%
Bosnia	BiH	3.8	2.9	76	51.2	37.3	73	1,087	3.0
Bulgaria	BUL	8.3	3.9	47	111.0	47.0	42	1,227	1082.0
Croatia	CRO	4.8	3.2	67	56.5	34.4	61	4,267	3.6
Czech Rep.	CZE	10.3	2.8	27	78.9	21.1	27	5,050	8.5
Hungary	HUN	10.2	10.2	100	93.0	93.0	100	4,462	18.3
Moldavia	MOL	4.3	1.1	26	33.8	12.0	36	504	11.8
Romania	ROM	22.6	21.2	94	237.5	237.4	100	1,549	154.8
Slovakia	SVK	5.4	5.2	96	49.0	44.3	90	3,624	6.1
Slovenia	SLO	2.0	1.7	85	20.3	17.5	86	9,101	8.3
Ukraine	UKR	50.9	3.1	6	603.7	32.4	5	976	16.0
Yugoslavia	YUG	10.4	9.0	87	102.2	88.9	87	1,462	18.5
Germany	GER	82.1	9.1	11	356.8	56.2	16	25,606	1.8
Austria	AUT	8.1	7.7	95	83.9	80.5	96	24,691	1.3
Total/avg.		223.2	81.2	36	1,878	802.3	43	11,532	

Source: R. Wanninger, *Socio-economic effects of water pollution in the Danube River Basin: Summary Report, June 1999*, Danube Pollution Reduction Programme, UNDP/GEF (United Nations Development Program/Global Environmental Facility); International Commission for the Protection of the Danube River, *Joint Action Programme for the Danube River Basin*. URL: www.icpdr.org

According to the data in Table 11.1, the DRB countries can be divided into three groups: from the "wealthy countries" (GER, AUT) in the upper Danube, through the "medium countries" (CZE, SVK, HUN, SLO, CRO) in the middle Danube, down to the "poorer countries" (BiH, YUG, BUL, ROM, MOL, UKR) in the lower part of the river Danube.

Per capita GDP in the thirteen DRB countries varies between USD 512 (MOL) and USD 25,600 (GER), or by a factor of about 50. The economic status of all the Danube countries taken together is documented by average GDP of USD 11,532 per annum. Yet among the formerly communist countries in the DRB (all except GER and AUT) average economic power is extraordinarily low—USD 2,046 per capita per annum, which classifies these Danubian countries as developing countries.

The composition of GDP by main economic sectors differs broadly from country to country (see Table 11.2). In the countries with the higher development levels (GER, AUT, CZE, HUN, SVK, SLO), the share held by the agricultural sector (usually including forestry and fishery) varies between 1 and 5.3 percent. As the economic power parity of the country decreases, the share of the agricultural sector in the GDP increases. The share of GDP generated by the industrial and tertiary sectors of the Danube countries is relatively comparable. With the exceptions of Bosnia and Herzegovina and Moldova, similar proportions of inhabitants reside in urban or rural areas of the DRB countries (Table 11.2). Also roughly equal are the population densities of these countries.

All of these geographic, demographic, and economic parameters are keys to understanding the problem of water management in the individual Danube countries. The demand for and quality of drinking water, the status of sewage systems, and the quality and quantity of wastewater must be considered the prime determinants of correct measures in water management in the DRB countries.

Drinking Water Supply

Numerous criteria are used to describe the current situation of the drinking water supply in the DRB countries. Table 11.3 presents some of the more important water supply parameters from the countries studied. "Water consumption" is restrictively defined as the quantity of water that is actually used by private households, which is metered and has to be paid for. "Water demand" is defined in this context as the quantity of water that has to be supplied to cover domestic demand, thus usually including consumption by private households; commercial, institutional, and tourism consumption; and losses in water production and distribution.

Domestic water consumption ranges from 98 l/cap/d (in CZE at the far low end of consumption) to 244 l/cap/d (extremely high consumption in ROM, probably due to agriculture). The rest of the countries have relatively comparable values. The cost of water plays a very important role in household water con-

Table 11.2 Population and GDP in the DRB Countries

	BiH	BUL	CRO	CZE	HUN	MOL	ROM	SVK	SLO	UKR	YUG	GER	AUT	Total
Present population of DRB countries														
Urban (%)	80	70	55	60	63	27	55	50	54	45	52	—	—	58
Rural (%)	20	30	45	40	37	73	45	50	46	55	48	—	—	42
Density (cap/km^2)	79	84	94	131	109	91	89	116	99	95	101	162	96	101
% of GDP produced by sector														
Agriculture	—	11.7	10.3	5.0	3.0	30.0	34.2	5.3	5.2	17.8	19.9	1.1	2.1	
Industry, mining	—	28.3	20.3	33.8	30.3	25.0	19.1	27.0	36.1	44.8	37.8	31.9	27.6	
Services, other	—	60.0	69.4	61.2	66.7	45.0	46.7	67.7	58.8	37.4	42.3	67.0	70.3	

Source: R. Wanninger, *Socio-economic effects of water pollution in the Danube River Basin: Summary Report*, June 1999, Danube Pollution Reduction Programme, UNDP/GEF; International Commission for the Protection of the Danube River, *Joint Action Programme for the Danube River Basin*. URL: www.icpdr.org

Table 11.3 Basic Characteristics of the Drinking Water Supply in DRB Countries

	BiH	BUL	CRO	CZE	HUN	MOL	ROM	SVK	SLO	UKR	YUG	GER	AUT
Domestic water consumption (l/cap/d)	150	190	170	98	107	143	244	131	141	144	179	146	145
Total water demand (l/cap/d)	250	439	254	265	147	177	409	245	196	172	255	230	242
Population connected to central water supply systems (%)	57	98	62	86	96	29	61	82	81	70	45	98	86
Range of losses in the water supply system (%)	40	43	35	28	27	20	22	23	28	17	30	12	13

Source: International Commission for the Protection of the Danube River, *Joint Action Programme for the Danube River Basin*. URL: www.icpdr.org.

sumption. Excluding Germany and Austria, water prices are still significantly lower (0.1–0.3 USD/m^3) than in Western countries although they have been rising. A decrease in total water demand and domestic water consumption has been observed over the last ten years in all postcommunist countries, mainly as the result of increasing water costs. This fact is documented in the example of the Slovak Republic in Figure 11.2.

The share of the national population connected to central water supply systems ranges from 29 percent (MOL) to 98 percent (GER). A connection value in the 60th percentile indicates the developing infrastructure of the country. Values above 80 percent are comparable with those of developed Western countries (the reported value of 98 percent in BUL seems to be a fabrication). Accurate data on water losses in the water supply systems are difficult to obtain; adjusted values are likely to be significantly higher in some DRB countries (UKR, MOL).

Wastewater Production and Sewerage

The basic characteristics of the production and treatment of municipal wastewater are presented in Table 11.4. From this point of view there are significant differences between individual countries. Regarding domestic wastewater production, there are no reliable data on the wastewater generation by population using individual water supply sources. The average wastewater production per capita is usually unknown (estimates range from 30 to 100 l/cap/day). The striking differences are in the figures reported for municipal wastewater that is channeled into central sewage systems in the DRB countries—from 80 l/cap/day (CZE) to 202 l/cap/day (SVK). These differences probably are the result of diverse calculation of production (inclusion or exclusion of infiltration to sewers,

Figure 11.2 Development of Domestic Water Consumption in the Slovak Republic (l/cap/d, 1990–2000)

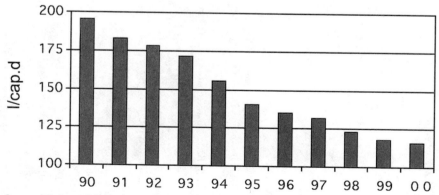

Source: Ministry of Agriculture of the Slovak Republic, *Report on Water Management in the Slovak Republic 2002* (Bratislava, November 2002).

rainwater, industrial wastewater, etc.). The next problem is the sewage quality. A high level of groundwater infiltration dilutes wastewater and lowers its temperature, causing problems in wastewater treatment. This circumstance is typical for all postcommunist DRB countries.

The fraction of the population connected to central sewage systems is relatively similar (40–60 percent) for the majority of DRB countries. The only exceptions are Moldova (14 percent) and Yugoslavia (33 percent), whose figures correspond to the level of economic development in these countries. A high degree of connection to sewage systems is characteristic of the Western countries Germany (89 percent) and Austria (75 percent), followed by Czech Republik (71 percent) and Bulgaria (65 percent). To provide a wider perspective of DRB countries with regard to connection to central sewage systems, the data of other countries in the world are presented in Table 11.5.

The economic problems following the downfall of communism slowed the development and construction of treatment systems in comparison with the previous era. Especially crippling were the financial problems of the new economies and the obscurity of the economic relations (e.g., privatization of utilities) they maintained. A typical example is the Slovak Republic (see Figure 11.3).

Municipal Wastewater Treatment

The different status of wastewater treatment in each of the DRB countries is the result of such factors as diverse historical and economic development, wastewater management traditions, and access to the sea. One of the relevant comparable indicators is the portion of untreated wastewater (directly discharged into the river system) in the total collected wastewater production for each country. From this point of view only the postcommunist countries Czech Republik and Slovak Republic achieve the best values (of course, GER and AUT do, also). In the countries recently stricken by the Balkan War (BiH, YUG and CRO), the fraction of untreated wastewater, shockingly, exceeds 80 percent. Meanwhile, despite their relatively strong economies, surprisingly high portions of wastewater go untreated in Slovenia (77 percent) and Hungary (58 percent). In Slovenia, of the total wastewater produced only 46 percent is collected and only 23 percent of collected wastewater is biologically treated; i.e., only about 10 percent of produced municipal wastewater is biologically treated (for HUN this figure is 23 percent).

The values presented above amount to a daily flow of about 6.5 million m³ of untreated wastewater into the Danube River system. Assuming an average municipal wastewater concentration of 200 mg BOD_5/l, they represent a BOD-load (Biological Oxygen Demand) of about 1.3 million kg BOD_5/d and about 21.7 million PE (population equivalent).[1] The prevailing part of this pollution is produced and discharged into the Danube in the territory of Romania.

In all DRB countries there is a strong tradition of using the activated sludge process in secondary wastewater treatment. The age of wastewater treatment plants

Table 11.4 Production and Treatment of Wastewater in the DRB Countries

	BiH	BUL	CRO	CZE	HUN	MOL	ROM	SVK	SLO	UKR	YUG	GER	AUT
Municipal wastewater production (l/cap/d)	125	161	178	80	139	152	197	202	108	157	140	155	145
Population connected to central sewage system (%)	52	65	41	71	45	14	41	50	46	51	33	89	75
Type of wastewater treatment (%)													
mechanical	0	—	13	3	44	0	43	5	37	0	8	0	0
biological/other	15	—	6	88	42	87	15	89	24	40	6	100	100
without treatment	85	—	81	9	14	13	31	6	40	60	86	0	0

Source: R. Wanninger, *Socio-economic effects of water pollution in the Danube River Basin: Summary Report, June 1999*, Danube Pollution Reduction Programmme, UNDP/GEF; International Commission for the Protection of the Danube River, Joint Action Programme for the Danube River Basin. URL: www.icpdr.org

Table 11.5 Connection to Sewage System in Selected OECD Countries (year 2000)

Country	CAN	MEX	USA	JPN	KOR	NZL	BEL	TUR	DNK	GRC	POL	NOR	ESP	UK	ITA
Population connected to central sewage systems (%)	78	22	71	62	53	80	27	12	87	45	47	73	48	88	61

Source: OECD: Environmental Data Compendium. URL: www.oecd.org

Figure 11.3 Development of Inhabitants' Connection to Sewage Systems in the Slovak Republic

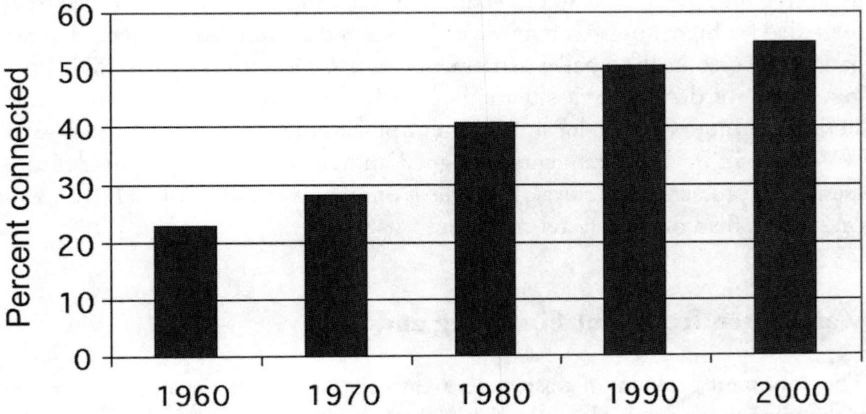

Source: Ministry of Agriculture of the Slovak Republic, *Report on Water Management in the Slovak Republic 2002* (Bratislava, November 2002).

is relatively high; more than half of them were constructed in the 1970s. There are practically no plants providing nitrogen and phosphorus removal. The lack of adequate industrial pretreatment is a problem in all of the countries. In some cases, industrial wastes cause BOD and COD (Chemical Oxygen Demand) influent concentrations that are much higher than in typical municipal wastewater. Industrial wastewater often contains toxic or other undesirable components: heavy metals, oils, toxic organic compounds, wastes, and so on. There are many wastewater treatment plants that are overloaded by 100 percent or more. Upgrading these facilities is an important strategic consideration for the short term.[2]

There is a lack of comparable data on types of wastewater treatment plants (WWTPs), the efficiency of WWTPs, and effluent parameters in individual DRB countries. At the highest level of the wastewater treatment, in Germany and Austria, WWTPs achieve high removal efficiency, including the processes for nitrogen and phosphorus removal. All the WWTPs above 100,000 PE fulfill the highest effluent criteria. The Vienna WWTP is currently being rebuilt and upgraded to the state of the art. Among the DRB countries, relatively high wastewater treatment status is attained in the Czech Republic. Having started on a relatively good level after the communist regime's downfall in 1989 thanks to its historic wastewater treatment traditions, research, and education in this field, the Czech Republic rose to first place in wastewater treatment development among all the postcommunist countries. Czech cooperation with Germany in recent years in the revitalization of the river Elbe has allowed the expansion and reconstruction of many large WWTPs in the Czech Republic. The Prague WWTP has waited for its upgrading for some years.

Of the total 2,883 Slovak settlements, only 471 have a sewage system; 363 have their own WWTP. The structure of WWTP distribution in Slovak settlements is demonstrated in Figure 11.4. It shows that a significant part (96 percent) of the population living in settlements with more than 10,000 inhabitants (72 in all) is connected to a WWTP. Many of the existing larger WWTPs need to be upgraded for higher organic removal efficiency and for nutrient removal. The situation is worse in the smaller settlements, where the connection to WWTP is low. Given the demographic situation of the Slovak population (see Figure 11.4), along with more realistic local investment possibilities, it is expected that small WWTPs will be most frequently designed and constructed in the near future. Despite its problems, the Slovak Republic is one of the best-developed DRB countries in the field of wastewater treatment.

Wastewater Treatment Financing and Policy

The dominant problem of wastewater treatment in all postcommunist countries is the investment costs. The lack of their own investment sources negatively influences these countries' development of water and wastewater management. Many world and European programs have been started (under UNDP, PHARE, ISPA, EBRD, the World Bank, etc.) with the aim of providing financial support for postcommunist countries in the field of water and wastewater management. The Environmental Program for the Danube River Basin (EPDRB) is one of them. Within the framework of the EPDRB, the costs of achieving the priorities of the individual Danubian countries for municipal wastewater treatment strategies were estimated. The expected reduction of pollution from municipal WWTPs is presented in Table 11.6. Expected total financial outlays for solving only the

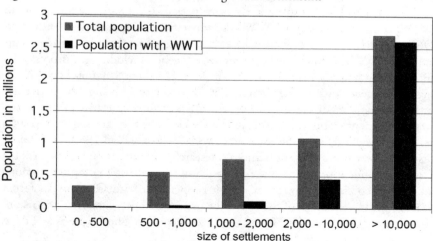

Figure 11.4 Distribution of WWTPs among Slovak Settlements

Source: I. Bodík and E. Rajczyková, "Basic Wastewater Treatments Problems in the Slovak Republic," *XXI. storočie* 4, no. 1 (2001): 30–32.

Table 11.6 Expected Reduction of Pollution from (tons/year) and Expected Investment Cost in Municipal WWTPs in the DRB for Years 2000–2005.

DRB country	BOD_5	COD	N_{total}	Investment costs (in millions USD)	The most important country investment action (in millions USD)
BiH	>7,690	>14,800	>3,000	147	WWTP Tuzla – 58
BUL	>19,448	>34,718	>2,308	>112	WWTP Sofia – 26
CRO	>15,310	>34,426	>1,509	434	WWTP Zagreb – 256
CZE	1,394	>597	1,306	208	WWTP Brno – 50
HUN	—	—	3,282	655	WWTP Budapest – 407
MOL	249	459	785	297	
ROM	24,121	27,274	5,167	597	WWTP Bucharest – 492
SVK	12,968	25,459	2,001	103	WWTP Kosice – 20
SLO	27,836	45,440	5,053	383	WWTP Ljublana – 110
UKR	>678	>621	>486	>55	WWTP Uzghorod – 25
YUG	—	—	—	—	
AUT	14,000	30,000	9,500	730	WWTP Vienna – 200
GER	1	9	1,700	110	WWTP Munich – 85
Total	>124,000	>214,000	>36,100	>3,850	

Source: International Commission for the Protection of the Danube River, *Joint Action Programme for the Danube River Basin.* URL: www.icpdr.org.

most critical national municipal wastewater problems are about USD 4 billion. But, on the other hand, the total outlays in the DRB countries for municipal WWTPs are much higher. For example, to meet all the requirements of European wastewater legislation just in the Slovak municipal WWTPs, about USD 2–2.5 billion is needed.

In principle, countries in the Danube region could borrow money for water pollution control. However, such a policy would not be wise, as the national debt is already high (between 30 and 80 percent of GDP). Some of the Western countries spent about 1 percent of their GDP on developing sewerage and wastewater treatment over the past twenty years and achieved remarkable results. To realize a similar development path, postcommunist countries would have to utilize at least 5 percent of GDP for the same purpose—an unlikely option.

One of the most important steps for the future development of water management (and the whole environment, too) is to develop a new long-term control policy. In the majority of the DRB countries the leading water management authority has not been the Ministry of Environment but other ministries: of construction, transport, industry, forest, etc. The allocation of legal competence among the state, district or municipal, and community levels varies dramatically from one DRB country to another. It usually depends on historical features and especially on the federal structure of the particular country. Responsibility for water supply and treatment and ownership of the infrastructure is being transferred to municipalities. Decision making has been largely decentralized. Imposed in the absence of experience or adequate institutional structures, decentralization can lead to rather peculiar schemes, particularly if financing issues also are involved. For example, the planned construction of a wastewater treatment plant may now involve decision making by the state and local governments, several ministries, their inspectorates and regional authorities, water works, and so on.[3]

European Legislation

The EU Urban Wastewater Treatment Directive (91/271/EEC), which was adopted in 1991, provides the main legislation for the control of urban pollution. The aim of the directive is to avoid pollution of fresh and marine waters by urban sewerage systems. The directive requires that:

- All agglomerations above 2000 PE should be provided with collection systems for urban wastewater;
- The effluent from sewage treatment plants must meet certain minimum effluent standards as laid down in the directive (the standards depend on the degree of sensitivity of the receiving body of water);
- Sewage discharges into "less sensitive" waters, which are defined as estuarine and coastal waters with high dispersion capacity, may receive only primary treatment;
- Sewage discharges into a body of water with "normal" sensitivity must receive at least biological treatment;
- Sewage discharges of > 10,000 PE into "sensitive" waters must be subjected to nutrient removal in addition to biological treatment; "sensitive areas" are to be identified based mainly on the risk of eutrophication.[4]

In 2000, the European Commission published a landmark document in the field of water policy, the Water Framework Directive (WFD). The overall purpose of the WFD is to establish a framework for the protection of fresh water, estuaries, coastal waters and ground water in the EU. It requires member states to identify the individual river basins lying within their territories and to assign competent responsible authorities.[5]

DRB countries in which the legal framework for environmental management is regarded as adequate and consistent with international requirements include Germany and Austria and, with some reservations, Hungary, the Czech Republic, and the Slovak Republic. In the other countries essential deficits and problems persist, mainly because in some countries the environmental and water-related legislation is based to a certain extent on historical structures and now suffers from critical inconsistencies brought on by the various changes, adjustments, and modifications. The majority of the DRB countries are currently in the process of establishing new environmental and water-related laws whose practical applicability and effectiveness have not yet been proven.

Thus, in most of the DRB countries the relevant legislation is currently in a phase of substantial reform and modernization. Due to the complexity of this task, it can be anticipated that the ongoing reform process and relevant legislation will take several years to complete, i.e., to reach an acceptable level of compliance with international requirements. In the context of this long-term accession process many postcommunist countries have adopted the provisions of "Chapter 22—the Environment" (within the framework of the EU accession process) but have negotiated transitional periods for urban wastewater treatment until 2010 (CZE) or 2015 (HUN, SVK, SLO). Bulgaria and Romania have not yet adopted the environment chapter of the *Acquis communautaire*.

Conclusion

This chapter has addressed the actual status of water and wastewater management in the postcommunist European countries lying in the Danube River basin. From the viewpoint of economics and development, these countries are quite differently positioned, but all of them have the same aim—the improvement of water quality in the Danube River basin. The DRB postcommunist countries are going through unique political, economic, and social changes associated with the heritage of serious pollution problems from the past. The solution of these problems requires tremendous outlays that are not in harmony with the resources available.

The overall level of water supply is quite high, and sewage collection is, on average, adequately developed in DRB municipalities. However, the quality of municipal wastewater treatment is poor. The choice of an optimal wastewater management strategy is very important. Could the historical development of wastewater treatment in Western Europe during last twenty years serve as the optimal model for Central and Eastern Europe?

The future of wastewater collection, treatment, and sludge disposal in the DRB countries ought to be focused on these main developments:

1. **Adopting legislation in the DRB countries** comparable to EU legislation in the field of sewerage systems, wastewater treatment, and sludge processing. It should focus on improved requirements for the quality of effluents from WWTPs in all monitored parameters, above all in nutrient concentrations. Standards in the DRB countries are currently under revision, and the application of a flexible system is being considered in order to set realistic limit values with respect to the recent economic and aquatic environmental situations.

2. **Implementing modern wastewater treatment technologies.** In the context of the need for WWTP reconstruction, it is necessary to utilize modern, low-cost technologies (e.g., use of biofilms), regulation systems of the WWTPs, etc.
3. **Monitoring and controlling** the flow rate and quantity of industrial wastewater discharged into municipal sewer systems.
4. **Building small WWTPs of under 10,000 PE.** In the DRB countries a relatively high amount of the population lives in settlements smaller than 10,000 inhabitants, where the shortcomings of the sewage systems and WWTPs are evident. For these settlements, low-cost technologies appropriate for small WWTPs must be used.
5. **Reducing the impact of poor design** based on incorrect data and repairing building defects in existing sewerage and wastewater treatment plants. Solution of these problems may help to decrease the inflow to treatment plants, but this requires huge investments.
6. **Addressing the problem of sludge processing and disposal.** The DRB countries will face some problems with sludge disposal. Therefore, all available modern and ecologically friendly methods for sludge processing must be exploited with the aim of minimizing the quantity of produced sludge and maximizing the exploitation of sludge in other industrial fields.
7. **Upgrading the skills of operators and the monitoring of WWTPs.** Many plants have been expanded or upgraded in several steps, resulting in complicated treatment lines, yet the standard of flow measurement and quality have not been changed. Operators' knowledge often is insufficient and behind the times with regard to technological developments.

Notes

1. One population equivalent (PE) means the organic biodegradable load having a five-day biochemical oxygen demand (BOD_5) of 60 grams of oxygen per day. Definition taken from EEA information website.
2. J. Námer, M. Drtil, I. Bodík, and M. Hutňan, "Wastewater treatment in Slovak Republic," *Polish Journal of Environmental Studies* 6, no. 2 (1997): 39–45.
3. L. Somlyódy, "Quo vadis Water Quality Management in Central and Eastern Europe," *Water Science Technology* 30, no. 5 (1994): 1–14.
4. Council Directive 91/271/EEC on Urban Wastewater Treatment.
5. Water Framework Directive, Directive 2000/60/EC of the European Parliament and of the Council, 23 October 2000.

Part Four

Agriculture and Rural Development

Chapter 12

"THINKING UNLIKE A MOUNTAIN"

Environment, Agriculture, and Sustainability in the Carpathians

Anthony J. Amato[1]

In *A Sand County Almanac,* conservationist Aldo Leopold described the Carpathian Mountains as one of the last islands of wilderness in Europe.[2] In his enthusiasm for the meadows and forests of a land he never saw, Leopold made a mistake similar to the one that he made in his reading of the landscapes of the southwestern United States, which he knew well.[3] In dubbing both areas wilderness, the conservationist missed a long history of human habitation and alteration in both environments. He took places that were made and managed for places that were untouched and wild. Although his description of the Carpathians did not correspond to the realities of the mountains in the twentieth century, Leopold's land ethic and his concern for knowing place have particular relevance for the Carpathians and neighboring regions as they enter the twenty-first century.

In the years that have passed since the fall of the Berlin Wall, environmental problems have emerged as some of the most pressing concerns in Central and Eastern Europe. Beyond the immediate problems created by polluting heavy industry, the countries of Central and Eastern Europe face larger questions about economic development, land use, and conservation. Those interested in these questions and in the problematic concept of sustainability can gain insights by examining the landscapes in one stretch of the Carpathians, the Galician Hutsul region. A study of this region, once part of the Austrian crown land of Galicia, demonstrates that the human and natural elements in a landscape are sometimes indistinguishable, and that inhabitants and nature in the mountains are intertwined in a host of shifting exchanges and relationships.[4] Despite the immensity of the mountains and their wild appearance, their landscapes, far from pristine, have been formed by a lengthy history of natural and cultural co-creation. Over the centuries, Hutsul shepherds and cultivators have adapted to the demands of the mountains, while at the same time changing local flora and fauna, nutrient cycling, and succession sequences to meet their needs. Nature has also proven to be an important actor, exhibiting a wildness prior to, during, and even after the socialist period. Therefore, finding a path of sustainable development in the Hutsul Region and other bioregions of Central Europe will require programs that both draw on inhabitants' "captured knowledge" and recognize the wildness of nature.[5]

In the Hutsul Region, the chain of mountains known as the Carpathians rises 1,800 meters above the lowlands of southwestern Ukraine and culminates in a 2,000-meter-tall ridge of peaks known as Chornohora (Black Mountain). Responsible for variations in temperature, precipitation, and slope, the mountains exert considerable influence on life and livelihood. Elevation, which determines the boundaries of ecosystems, forms the basis for six distinct bands of natural and cultural landcover in the Hutsul Region. The first band, which reaches up to 650 meters above sea level, supports grain and contains some alpine flora; the second band stretches up to 950 meters above sea level, where wheat can no longer be grown effectively and 50 percent of the flora found in the valleys disappears; the third band (950 m to 1,250 m above sea level) is covered by a belt of beech trees and evergreens; the fourth band, which starts at 1,250 meters, supports only evergreens; the fifth band (1,600 m to 1,800 m) is characterized by a mixture of mountain pines (*zherepy* in Ukrainian) and alpine meadows; finally, in the sixth band (1,800m+) the mountain pines thin out, leaving only alpine meadows. Within these bands, 950 meters above sea level stands as the upper limit for wheat cultivation. Other crops have different limits. Corn (maize) thrives at elevations up to 800 meters, rye grows at 1,100 meters, and barley and oats can be sown at 1,200 meters. The cultivation of potatoes extends all the way up to 1,250 meters.[6] Although settlements range in elevation from 350 meters to 1,700 meters, the vast majority of people in the Hutsul Region live within the first two vegetational bands (below 950 meters), and even in the twenty-first century, most villages are still located not far above 350 meters above sea level, the dividing line between the mountains and the foothills.[7]

Relief, elevation, and its recent settlement have placed the Hutsul Region at one end of the spectrum of Galician agriculture. In the early twentieth century, only 4 percent of the region's surface was classified as tilled fields, and even villages with the largest percentage of tillage had limited land available for cultivation.[8] For example, in 1931, the village of Krasnoïlia, located in the wide Zhab'ie-Seliatyn Depression, had an above average percentage of tilled land (7.2 percent), but villages located in steep, narrow valleys had less tillable land: in 1931, the village of Kryvorivnia consisted of 2.2 percent tillable land. Locations deeper in the mountains had almost no land usable for fields and gardens. The commune of Hryniava, most of which was alpine pasture, had only 0.5 percent tillable land.[9] According to one estimate in the 1890s, the cultivated fields in the region could not feed even a tenth of the local population.[10] The lack of arable land has meant that the population density in the southeastern mountains has been and remains the lowest in Galicia. In the early twentieth century, the Hutsul Beskyd Mountains and their foothills averaged only 34 inhabitants per square kilometer, with their southern portion inhabited by a scant 10 people per square kilometer.[11]

Faced with limited prospects for cultivating cereals and garden vegetables, inhabitants have relied heavily on the mountains' forests and meadows. Prior to 1939, mountaineers counted on the forests for fuel and building materials, and they depended on the sale of timber and on employment in the logging industry as important sources of income. They made the undergrowth and the forest floor important parts of their world: here they pastured their livestock, collected herbs and mushrooms, and picked blackberries, blueberries, raspberries, and currants. The forest sustained life in the mountains.

Like the forest, the "alps" or "high pastures" (*polonyna*—Ukr.) are an integral part of the Carpathians.[12] Situated above the timberline and covering large stretches of the mountains, the *polonyna* is a complicated collection of ecosystems. Alpine meadows contain a range of flowering plants, including orchids, gentians *(Gentiana lutea, Gentiana punctata)*, and arnica *(Arnica montana)*. They also produce a range of "lower quality" fodder plants, including hard rush *(Juncus trifidus)*, sedge *(Carex curvula* and *Carex ruperstris)*, matgrass *(Nardus stricta)*, and *Avena* and *Deschampsia caespistosa*. The alpine habitat is home to shrubs and bushes as well, with mountain alder *(Alnus viridis)*, rhododendron, and junipers sprinkled across the meadows.[13] Most alps border evergreen forests, which mountaineers used for additional livestock forage and for materials to build fences, sheds, and huts. The vast alpine pasture and the thick forests below them provide two sharply contrasting but interlinked landscapes.

The *polonyna* formed the basis for alpine husbandry, which was crucial in the Hutsuls' pre-1939 ecological system. Every summer, groups of men herded their villages' sheep, goats, cattle, horses, and pigs up to the high pastures. The summer-long treks began when the snow melted on Chornohora (in late May or June) and they ended in late September. Shepherds spent their four months on the *polonyna* milking the animals and making cheese to send back to the village or to market. The summer treks to the uplands were crucial. The *polonyna* enabled the Hutsuls' animal husbandry to function in two different sets of ecosystems and under two different economic arrangements. Villagers pooled their animals for the summer, and flocks expanded beyond the limits placed on them by village pastures and meadows. Transhumance permitted mountaineers to maintain a pastoral economy larger than the one that local resources alone could support.[14]

The region's lands, their agricultural possibilities, and inhabitants' activities have led to peculiar farm and village layouts. Unlike many other Europeans who live in small, compact villages surrounded by fields, Hutsuls live apart from one another, sometimes at distances of half a kilometer, and within the boundaries of a given village, elevation varies by several hundred meters or more.[15] In addition to the holdings adjacent to their houses, prior to 1939 families owned swatches of pasture and forest scattered throughout the commune. Settlement and landholding patterns have resulted in villages that extend in all directions, crossing natural boundaries and incorporating a variety of ecosystems.

The relationship between human and natural systems in the mountains was formed over "deep time." While the plains of Europe were composed largely of old soils *(Altland)*, originally cultivated by Neolithic agriculturists, the Carpathians and other European mountain ranges were composed almost exclusively of new soils *(Neuland)*, first cultivated during the Middle Ages or in more recent times. In medieval Europe, the Carpathians were a frontier where markets, migrations, and processes of ecological change came together. After 1000 A.D., settlers from Galicia in Rus' began to trickle into the mountains, settling alongside and mingling with groups of herdsmen who resided in the highlands. The settlement and agricultural transformation of the Carpathians accelerated greatly during the time of the Second Serfdom (1450 to 1700) in Central and Eastern Europe. During this period, waves of serfs seeking relief and better service arrangements migrated to the mountains from lowland areas located in modern-day Ukraine.[16]

Upon arrival in the mountains, the ancestors of today's Hutsuls struggled to make a living below the fog-shrouded peaks. In one of his sketches, village teacher Mykhailo Lomats'kyi emphasized the precarious nature of subsistence: "Here in the Hutsul Mountains even the small child knows that a person does not live by

Figure 12.1 A typical Hutsul village in the Carpathians. Photo by Anthony Amato and Felixa Amato.

bread [grain] alone."[17] As late as the middle of the nineteenth century, Hutsuls suffered frequent crop failures, which produced regular famines. In "hungry" years, mountaineers turned to their surroundings, consuming pigweed, ripple-grass, nettles, and mustard grass.[18] During one famine in 1859, children in the village of Zhab'ie had to resort to eating grass, and their parents were reduced to eating the new leaves of beech trees.[19] The yearly entries from a Greek Catholic pastor's chronicle paint an equally bleak picture of life in the mountains: "1786—a hard year: a great famine on the Hungarian side and in our mountains ... 1816—a hard year, great traveling. Hutsuls went all the way to Moldova for grain."[20] As the entry for 1816 illustrates, mountaineers were forced to rely on the lowlands for a substantial portion of their calories, and much of a household's labor was devoted to raising money to purchase its "necessary reserve of grain."[21]

To make a home in the Carpathians, Hutsuls imported biota that originated in distant locales. By the middle of the nineteenth century, two New World crops, corn (maize) and the potato, provided most of the mountaineers' calories. Introduced to Poland in the second half of the eighteenth century, the latter changed life in the mountains.[22] Because potatoes required less processing, produced more calories per acre, thrived in rocky soils, and grew at significantly higher elevations than grains, Hutsuls cultivated ever-more marginal lands, and settlements appeared in locations farther from the lowlands and higher up slopes.[23] The potato-driven expansion of the late eighteenth and early nineteenth century came to a halt in 1846, when blight struck the mountains.[24] Biological invasions, the unwelcome counterparts of importation, kept the mountains wild.

By the 1850s, the inhabitants of the mountains had worked out diverse ecological and economic systems to ensure their survival. In the nineteenth century, the holdings of a typical household included a garden of 0.25 to 1 hectare, as well as fields covering several hectares. Mountaineers planted a range of crops in their gardens: potatoes, beans, peas, cabbage, radishes, turnips, onions, garlic, and occasionally beets. They sowed their fields with corn, oats, barley, and buckwheat; and, in addition to these four grains, a few wealthier families planted wheat and rye. To supplement their gardens and fields, most families had a few sickly fruit trees, with pear, plum, apple, and cherry trees scattered thus throughout mountain villages.[25] Agriculture in the mountains was, and remains, diverse by necessity.

Inhabitants of the mountains devoted a substantial amount of their time and resources to the neverending task of converting their surroundings into a usable form. Up until the middle of the twentieth century, the Hutsuls, one of the last groups of swidden (slash-and-burn) cultivators in Europe, cleared forest land for agricultural use, employing a variety of approaches that revolved around a progression of operations and cultures.[26] Preparing wooded land for cultivation was laborious. Peasants first selected an area and then cut all of the large trees down. Having carried out their logging, they set fire to the undergrowth and fallen brush before planting. They referred to these newly cleared places, which were usually located alongside settlements, as *palenyky* or *palenyshcha* ("scorched places"). When cultivators finally planted, "depending on the right progression of cultures and the humus content of the soil, the length of exploitation lasted ... for one or two years, sometimes even three."[27] The first crop sown in swiddens was oats. The following year these fields were usually sown with rye or left for hay or pasture. In the late nineteenth century, the potato began to dominate new fields, often taking up the first and second years, and wheat increasingly joined rye and oats in progressions. Once the progression was completed, cultivators abandoned the field, returning to the same parcel to start the process over again any time from 17 to 100 years later.[28]

In the second half of the nineteenth century, Hutsul agriculture underwent changes that had far-reaching ecological and economic consequences. Prior to the 1860s, few mountaineers had invested much effort in cultivation, but in the final decades of the nineteenth century, inhabitants increasingly engaged in cultivation, ignoring their previous concerns.[29] Because cultivation had strict limits in the Hutsul region, the techniques associated with it remained quite simple. Even in the early twentieth century, a number of families still turned the earth with picks, spades, and hoes rather than plows. As the size of their fields increased, more cultivators turned to plows and animal power.[30] Mountaineers introduced equally important changes in crop rotations. Although some still employed long-fallow systems in the twentieth century, they were using these and other systems that involved deforestation with less and less frequency. In the late nineteenth century, many began to employ multi-crop rotation schemes, usually without a fallow field. Large numbers of cultivators began to sow wheat and rye in the spring, and later, they began to sow wheat and rye in the fall as well.[31] In their new cropping schemes, they introduced cover crops, such as clover and lupine, and they began to use manure, compost, and mineral fertilizers along with the traditional fertilizer of ash.[32] By the twentieth century, intensive land use had expanded and begun to displace *extensive* land use. There is ample evidence of the benefits that mountaineers derived from intensification.[33]

Intensive grain cultivation transformed the landscape of the mountains. Many Hutsuls plowed up sections of their meadows and sowed them with grain.[34] Over the course of the nineteenth century, cultivated areas grew in size and number, and fields claimed a larger and larger portion of villagers' lands. For example, in 1788, the village of Kosmach had no tillable land at all, but by 1900, the village had 158 hectares of tillable land.[35] The expansion of fields accelerated during the first three decades of the twentieth century, and by 1931, Kosmach had 349 hectares of tillable land. Statistics reveal that similar increases occurred across the Hutsul Region. In 1901, tilled fields comprised 4.9 percent of the land in the Kosiv district and 8.1 percent of the land in the Nadvirna district. By 1932, tilled fields made up 7.6 percent of the Kosiv district and 12.7 percent of the Nadvirna district.[36] The expansion of tillage and pasture came at the expense of forest. In the villages of Holovy, Krasnoïlia, and Kryvorivnia, over 90 percent of the land was densely forested in the seventeenth century, but in 1931, residents used approximately 75 percent of the surface area for agriculture and husbandry.[37] By the twentieth century, only one third of the area in the Zhab'ie-Seliatyn Depression and the Hutsul Beskyd Mountains was forested.[38]

Husbandry in the uplands also changed between 1848 and 1939. In the second half of the nineteenth century, mountaineers began to cut hay in the uplands once a year and take it down to their villages for the winter.[39] The suspension of summer treks during World War I brought about significant changes in transhumance, too. When shepherds returned to the high pasture after the war, they brought smaller flocks and fewer species to the high pasture, with sheep dominant.[40] Peasants also modified grazing on the *polonyna*. Throughout the second half of the nineteenth century and the early twentieth century, they divided alpine pastures into two halves, one half designated as a hayfield and the other as pasture.[41] In many instances, they would use the same half for grazing several years in a row and the shepherds' sheds and pens would stand on the same place. Continuous grazing had visible effects. By the early twentieth century, a number of shepherds noted the damage that had been done to alps: luxuriant grasses disappeared and inferior grasses took their place, and certain areas were left completely bare.[42] To guarantee the long-term health of the high pasture and allow for both grazing and hay, shepherds began to use a rotational scheme in the 1930s. Started by the chief shepherd Dmytro Dem'ian on the alpine pasture Dukon'ka Velyka in 1933, this system involved switching halves (hay to pasture) every year, moving animals to different parts of the pasture every three to seven days, and regularly rotating the site of pens and sheds.[43]

The introduction of this system was a major step in making alpine husbandry more productive and sustainable. Highland pastures had been subjected to many different intense grazing pressures, but after the 1930s, these pressures were reduced, spread across a larger area, and in part replaced by the demands of haymaking. This spared alps from the stresses of intense selective and nonselective grazing. Twentieth-century practices resulted in alpine meadows that were more uniform in their vegetation and nutrient cycling. Regular removal of material from all plants, along with the fertilizing of whole meadows in some places, created meadows different from those that had been marked by grazers' peculiar preferences, irregular trampling, and patchy deposition of manure. High pastures after the 1930s were different from both the pre-1800 lightly grazed pastures and the early twentieth-century overgrazed pastures.

The demands of distant markets affected the region's forests. For centuries, Galicia's lowlands had relied on the mountains for timber, and in the nineteenth century logging increased when railroads tightened the link between the mountains and markets.[44] In the 1880s, companies from outside the Hutsul Region began logging its slopes to provide Central Europe with wood.[45] In 1883 and 1884, ten sawmills were operating in the Kosiv District, and another seven were operating in the neighboring Nadvirna District. In 1899 and 1900, almost 7,500 rafts of logs floated down the Cheremosh River alone, and by the early twentieth century, six to eight million cubic meters of timber were floating down the Prut and Tysa Rivers every year.[46] Logging, which entailed the removal of a portion of the region's biomass, transformed the Carpathians. Mountainsides of stumps, branches, and bare earth appeared after clear-cutting had taken place. The number of species increased sharply on a local level immediately after the logging, but biodiversity plummeted over time as species vanished and many young and identical ecosystems came to dominate the landscape.

Both Hutsuls and outsiders created many features of nature as they knew it. Frequent human disturbances, including logging and fire, were responsible for many of the meadows below the alpine zone. Moreover, humans greatly changed the composition of the region's forests because anthropogenic disturbances favored evergreens and other fast-growing early-successional species over slow-growing and late-successional species. The vast stretches of bare trees, stumps, and scorched earth were not antithetical to the forests and meadows of the Hutsul Region. Rather, the columns of smoke and the smell of charred wood were necessary for the region's blankets of evergreens and the scent of pine.

Animal husbandry also changed the mountains. In forests, livestock disturbed the litter and intervened in decomposition, preventing nutrients from reaching the soil. They introduced damaging fungal diseases and they consumed and trampled many seedlings and bushes, eliminating whole species in places.[47] Animals also changed floral populations and floral diversity in meadows. As many mountaineers pointed out, their animals generally preferred flowering plants and avoided tough, woody plants. Years of grazing left meadows with stiff, bitter, and bristly grasses, including sedges, hard rush, coltsfoot, sheep fescue, and smallreed. Flowering plants, including arnica, gentians, and orchids, gradually disappeared from pasture land.[48] At lower elevations, intensive grazing triggered a succession whereby shrubs and trees took over grassland.[49] Human attempts to improve and restore pastures, in some cases, did further harm to flowering plant species. Interwar efforts to fertilize pastures and meadows with manure encouraged the growth of tall grasses, which choked out low-lying plants such as edelweiss.[50] Many alpine plants, which faced little competition in nutrient-poor soils, could not compete after humans leveled the playing field by making alpine soils more fertile.

After two centuries of human and natural action, a new flora took hold in the mountains. Both forests and meadows were changed by exotic plants. Often introduced from seeds in the manure of grazing animals, these plants quickly took hold, finding or defining niches. The rhododendron and rye grass were two of the most successful species to spread across the landscape. These and many other exotics spread so widely that several decades after their introduction they were assumed to be native species.

Human actions and preferences affected the populations of native species, too. Mountaineers' preference for maple and oak for crafts and building resulted in a large shift in tree populations. Over the course of the nineteenth and twentieth centuries, the percentage of conifers in forests rose considerably, and nearly all stands of maple and oak that were to be found were young.[51] Similarly, beech forests, which once covered 55 percent of the Ukrainian Carpathians, accounted for only 33 percent in the modern period.[52] Humans found the wood in beech forests and the rich humus layer beneath them useful, and because of the beech tree's slow growth and its place in successions, stands of much faster-growing species grew in place of beeches when forests began to regenerate.

In addition to changing local vegetation, the expansion of agriculture, animal husbandry, and forestry affected the fauna of the mountains. Increasing pressure from humans reduced some animal populations. Deer and European bison dwindled in numbers as Hutsuls' rifles took their toll. The traditional large predators of the Carpathians, the wolf and the bear, were targeted by peasants, who saw them as menaces to their herds. At the same time, humans introduced and supported a host of animals inadvertently. Mountaineers laid the groundwork for the successful entry of sparrows, swallows, moles, mice, and rats. They helped these species gain a hold by eliminating predators and creating habitat and a food supply for the new arrivals.[53] Slowly but surely, humans changed local animal communities.

By the end of the nineteenth century, humans were quite conscious of their ability to alter the environment of the mountains.[54] Locals and outside foresters were so aware of changes in their surroundings that some of them initiated strategic tree-replanting efforts.[55] Like the replanting efforts launched by other Europeans, these efforts were influenced by humans' sense of productivity and resources. Smitten with the "fir and spruce mania" that was sweeping Europe, Carpathian foresters planted *Picea abies* widely, displacing beech and oak trees from many of their former ranges.[56] As part of their efforts, they also introduced exotic conifers to the Hutsul Region, including larches and the Douglas fir.[57] By the 1990s, the latter, native to western North America, dominated many slopes and crests. Through a century of deliberate tree replanting and other efforts to manage the timber of the mountains, humans had created new forests.

The human impact on the alpine landscape is more than just skin deep. Profound unseen changes have occurred in the region's forests, meadows, and fields. Although many past observers quickly identified invading and vanishing species, only recently have observers become aware of transformations in nutrient concentrations, soil composition, and biogeochemical cycles. The ecological changes set in motion by intensive grain cultivation went beyond landcover. The introduction of crops and fertilizers changed nutrient cycles. Deeper and more frequent plowing reduced soil moisture. By maintaining fields in continual rotation, mountaineers kept large stretches of land in a state of constant disturbance, and they put nutrients into the soil only to remove them when they harvested the plant matter with the taken-up nutrients. Along with the obvious changes in landcover that it entailed, the process of converting forests to fields and pasture shifted the largest concentrations of organic matter from an above-ground to a below-ground location. In logged ecosystems, above-ground nutrients greatly decreased in relation to soil nutrients, which increased once grasses, bushes, and other types of low-lying vegetation took hold. Deforestation also boosted the amount of oxygen and ammonium hydrate found in the soil.[58] Even forests that looked similar to the naked eye varied greatly. In one study of the upper Prut watershed, local old-growth mixed fir forests had a nitrogen exchange rate that was 1.5 times greater than that of a young fir-only forest. The study also revealed that the managed forest had five times less phytomass in it than did a similar old-growth area.[59] The vast differences between these forests suggest that humans have succeeded in replicating forms, but not functions or whole systems.

While humans were engaged in transforming their landscape, natural processes continued. Species moved, competed, and cooperated. This was evident in the region's seemingly still forests, where over time tree species took hold, rose to prominence, and then faded. These processes showed that nature itself was quite durable. In the opinion of two scientists, the human alterations that converted forest to pasture in the Prut Basin were not sufficiently severe or enduring to stop the regrowth of forest. Only in cases where erosion was severe did humans permanently alter the vegetation-succession regimen.[60] The regrowth of forest demonstrates that even the conversion of forest to pasture was not an act, but rather a continuous process that required regular grazing, mowing, and cutting to prevent succession. Although the forests of the Carpathians have not reached their pre-1800 boundaries, trees continue to take hold in neglected meadows.

Mountaineers' actions contributed to many unwelcome "natural" forces. Centuries of deforestation left the region vulnerable to rapid runoff and erosion. In the late nineteenth century, widespread clear-cutting stripped many slopes of vegetation, and on these slopes, otherwise ordinary and slow processes had the potential to change the landscape in sudden events. In an extreme—but not unique—case in June 1894, after late spring rains, a mudslide consisting of 10 square hectares of mountainside inundated a valley, damming a stream and creating a sizable lake in the process.[61] Greater and faster runoff also increased the frequency and magnitude of floods in the nineteenth and twentieth centuries.[62] Humans' contribution to these "natural" disasters in the mountains was clear early on. In the 1830s, an older mountaineer offered a sophisticated explanation for the more frequent flooding. He noted that deforestation had exposed snow packs to more sunlight, which caused it to melt faster and run off all at once.[63] In addition to exposing more snow to sunlight, deforestation meant the loss of water-absorbing soil, roots, plant cover, and litter layers.[64] Every year, rain washed fields away bit by bit, and swollen rivers carved deeper and deeper into valley pastures.[65] By the 1990s, tremendous spring floods in the Carpathians had become a yearly event, and residents seemed incapable of doing anything, except retreating and rebuilding. Nevertheless, the flooding implicated humans in many lasting and far-reaching changes, and demonstrated nature's agency and its relationship to humans. Much as humans had channeled and directed natural action and processes through means such as agriculture and water mills, nature could also bring together and channel disparate human actions, magnifying them and taking them in unintended directions.

Floods, forest regrowth, and exotic invasions demonstrate the degree to which mountaineers could alter their environment: they could add or remove some elements and intensify or stop some processes within ecosystems, but they lacked control over all elements and processes within these ecosystems. The result of this partial control was a landscape that was fully fashioned neither by humans in their own image nor by nature in its own image. Today the landscape of the Carpathians includes both an intended and an unintended environment.

Forces defined as cultural have also acted on the landscape of the mountains. Recent political and economic changes in Central and Eastern Europe have had a profound impact on agriculture and altered the environment of the Carpathians. World War II had particularly far-reaching consequences. In 1939, the Galician Hutsul Region was incorporated into the Ukrainian Soviet Socialist Republic. After the war, Soviet officials, determined to control all means of production, ordered the seizure and state control of all agricultural and forest land. In seizing property, the Soviet state imposed an ecological order along with the socioeconomic order it sought to impose. Private forests, alps, fields, and meadows passed from their owners' hands, and collective farms took over alpine husbandry and other large agricultural operations. Soviet managers maintained land-use patterns, but they expanded the scale of all operations and greatly increased the use of mechanization and chemical inputs.[66] Under the Soviet system, tillage continued to expand, with the land in cultivated fields in the Prut River basin increasing by 80 percent between 1949 and 1979.[67]

During the Soviet period, an unofficial agricultural system emerged in the mountains. The Soviet state had little success in making local agricultural lands productive, and as a result, it collectivized relatively little peasant land and many families continued to hold pastures and meadows *de facto*.[68] Not long after "total collectivization" had been achieved, the state allowed peasants to keep their own animals and gardens. Peasants directed much of their energy toward enlarging their own gardens, which incorporated a diverse array of new plants, including tomatoes and strawberries. In doing so, they further elevated the status of the garden as a productive unit in their agricultural operations. They also returned to keeping their own livestock. Although they were supposed to pay for these animals' grazing and hay, they routinely slipped them into collective farms' grazing herds, and in the absence of supervision, they moved them across forests and meadows. Cattle freely foraged in the dense pine forests on the steep mountainsides of villages such as Kosmach, making their presence known only with moos, snorts, and cowpies.[69] Throughout the Soviet period, grazing pressure in the region was far more widespread and pronounced than officials estimated, and meadows and pastures suffered from overgrazing caused by this second, or "shadow," husbandry.

From 1945 to 1989, peasants secured their place in the mountains. After the war, the Soviet state attempted to diversify the region's economic base by introducing light industry and continuing to develop tourism in the mountains. Both sectors provided locals with additional income, and locals engaged in wage labor across the Soviet Union and returned with small sums of money.[70] Because money varied in utility in the late Soviet period, Hutsuls found it advantageous to use their earnings to enlarge their agricultural operations. After paying for food and other expenses, they pumped their earnings into their operations, purchasing implements, seed, and animals. Household agricultural operations prospered and most villages grew.

While peasants were busy carving out space and securing their operations, Soviet officials emphasized conservation and rational husbandry in agriculture and forestry.[71] Against the odds, they managed to reverse processes that were centuries old. Many stretches of the mountains saw a return of forest. In the Prut River basin, forests, which had accounted for 72 percent of the basin's area in 1855 and 59 percent in 1949, covered 76 percent of the basin's area in 1979.[72] The Soviet state initiated environmental management and monitoring programs, and it also created conservation reserves and parks in the region, such as the Carpathian State National Nature Park in 1980 and neighboring Transcarpathia's Carpathian State Reserve in 1968.[73] In many cases, though, Soviet conservation served human needs, and in the case of forestry, it was oriented toward timber harvests or even harvests of herbs and berries. Soviet foresters, like many of their counterparts, insisted that management was too complicated and important to be left up to nature, and their restoration efforts focused on the rational over the natural. They concentrated on rehabilitating easily accessible areas and on maximizing timber production by planting forests with identical trees (often with a set distance between each). The Soviet legacy in alpine agriculture and forest management was a landscape of ecosystems that were low in biodiversity and adaptability but highly productive.[74]

By the 1980s, the long-term prospects of Soviet agriculture in the mountains were unsure. Yields had grown tremendously, but the mountains suffered from topsoil loss, damaged pastures, declining biodiversity, and pollution from fertilizers and biocides.[75] By 1989, Soviet agriculture was facing an even more immediate human and economic crisis. Operations were in disarray, the distribution and support systems were collapsing, and the agricultural labor force was dwindling. Household operations were facing problems of their own. The growth of small-holdings put pressure on resources. Families demanded wood for building houses. Their private herds and gardens grew and required more land. Even lesser known needs began to take a toll: for example, humans greatly reduced the numbers of some gentians considered medicinal herbs in the mountains.[76] Although they did not capture many headlines in 1989 and 1991, the environmental issues in the Carpathians were no less important than those in the headlines.

Tossed into the post-Soviet era, the Hutsul Region faces an uncertain future. The economic collapse of the post-Soviet world has changed the mountains. Without capital and markets and with the infrastructure collapsing, the ex-Soviet industries have abated their intense exploitation of the environment. While the pressure from official industry has abated, pressure from unofficial activities and individuals has greatly increased. Facing unremitting inflation and unemployment estimated at higher than 20 percent in some areas, households have turned to local sources of income.[77] To meet immediate needs, they have increased cultivation and grazing wherever they have had the opportunity. Many have taken advantage of the state's agricultural landholdings, which have been returned, leased, sold, or offered for free use to former employees, pensioners, former owners, and even interested locals. Households have also diversified their operations, and many are raising crops and animals that they have not raised in years. Households' reliance on local activities has drawbacks, though. Increased agricultural use threatens to undermine the health of local lands. Other trends threaten the mountains as well. With the collapse of the regulatory state, poaching and illegal logging have exploded.[78] In 1994, some locals were paying twenty dollars for large logs cut illegally, and many others paid nothing for the state-owned timber they took for building projects.[79]

At present, Hutsuls live in an era of great possibilities and perils. The last fifty years in particular demonstrate that they have created resilient systems of production in the Carpathians. Pre–World War II organizational forms, including pooled animals, head shepherds, and probationary milkings, have once again become the foundations for high pasture husbandry.[80] Hutsul agriculture is highly energy-efficient and biodiverse, and unlike many proposed alternatives, it requires few if any state subsidies. A unique intersection of economy and environment, it has outlasted three states and two economic systems, proving its durability. Creating a sustainable agricultural economy from this durable form will require considerable adjustment, however. The long-term increase of agricultural activity in the mountains poses environmental challenges, and many current perils stem from the slow processes that characterize interaction between humans and the environment. Both the present and the past demonstrate the power of cumulative effects of small actions over many years. Viewed in one place and at a single time, the results of the expansion of settlement, agriculture, and forestry in the Carpathians have been largely invisible; it is only when viewed over decades and centuries that the results of these trends become evident. When the creation of new fields and meadows is framed as part of a thousand-year trend, it is clear that Hutsuls have transformed their surroundings and placed more demands on their environment. In almost every village, the effects of damaging gradual processes are evident. For example, in Kosmach the banks of the Pistyn'ka River have become high and bare, which testifies to the recent history of flooding.[81] The recent floods and the accompanying loss of agricultural land in the region have shown that long before humans damage that which they define as nature, they can destroy much of what is human and cultural.

Other developments point to favorable outcomes for humans and nature in the mountains. The history of humans and environment in the Carpathians is not necessarily a linear progression. In the opinion of two scholars, human pressure on the mountains increased from the late Holocene to 1944, but after 1944, pressure on the Polish Carpathians subsided while pressure on the Ukrainian Carpathians, including the Hutsul Region, increased.[82] The divergent paths of these two areas testify to the many different types and levels of human action in shaping the landscape. Moreover, current population trends based on low fertility and out-migration have resulted in slow growth or slow decline in areas of the Hutsul Region.[83] These trends will give the region a respite from two centuries of increasing small-holder pressure, and at the same time, slow population growth in some places has prevented the population implosion that has undermined operations in other areas of Ukraine. Divergent paths, population trends, and the regrowth of forest during the Soviet period shows that humans' changes are rarely permanent, and the notion that extreme human pressures can subside and even be reversed offers hope.

Sustainability in the Carpathians requires considerable work and sacrifice. Residents must drastically alter their activities in many places to avoid losing what remains. They must identify, know, and protect thriving places. In their conservation efforts, those who seek to restore nature must avoid the temptation to focus on the mountains as a museum of wilderness or culture. The Carpathians have ample room for both nature and culture.[84] The complete removal of agriculture and forestry would likely cause biotic communities in the mountains to be diminished or imperiled, insofar as natural processes alone "account for very little of the total variation ... in ecosystems."[85] The absence of human action would also reduce or eliminate ecotones vital for many species, and as recent research suggests, even swidden cultivation, which integrates the wildness of fire, has a role in the mountains. Swidden disturbances created the forests of the Carpathians, and eliminating these disturbances creates fuel loads, not forests.[86] Both the uninterrupted and patchwork landscapes of the mountains are the products of thousands of years of human action on top of natural action, and the region's rich biodiversity and vital processes have depended on the Hutsuls' practices. Accepting human action in the mountains and seeing it as a part of nature is an important step toward a sustainable development that respects both the wildness of nature and agriculture's own long history.[87]

The past and present of the Hutsul Region offers lessons about wealth and health for its neighbors and for the Third World. Some have argued that the First World should encourage economic growth first and leave environmental protection to follow with the creation of wealth.[88] Nevertheless, fifty years of development in the Carpathians shows that development programs guarantee neither environmental health nor local wealth. In fact, some Soviet development programs in the Hutsul Region did great harm. Counseling countries to "develop out of" environmental problems will guarantee that the Carpathians and other locations in the new Central Europe "develop into" other problems, such as the tremendous waste that plagues consumer economies. Unmanaged growth also threatens one of the positive legacies of the Soviet state: large stretches of undeveloped and relatively untouched areas.[89]

Making good decisions about sustainable development is difficult. Residents of the Hutsul Region face a dilemma: they suffer from environmental degradation, but they also suffer from poverty and see the mountains' forests and meadows as valuable resources. Throughout the region, the immediate consumption of resources has an almost irresistible allure, and the tendency of immediate needs to outweigh long-term needs is manifest in everything from clear-cuts to trout poaching. In reviewing similar situations worldwide, some commentators have put forward a bold proposition: environmentally based development is the only route to wealth. They have argued that not only are the demands of nature and humans not antithetical, but environmental protection is actually a prerequisite for economic development. Conditions in the Carpathians suggests that these observers might be right. While the failure of environmental protection and economic development during the socialist period is attributed to a "Tragedy of the Commons," the failure of environmental protection and economic development in the postsocialist period is due to a "Tragedy of Enclosure."[90] Too much has been allowed to belong to too few, with too low a price being paid for it. In many parts of Central Europe, privatization and many forms of enterprise have amounted to small groups getting rich off of stolen property and destroying local capacities to produce wealth. In contrast, development based on stringent environmental protection laws and local programs requires users to invest, build, and use efficiently. By reasserting the commons and making the health of landscapes the first priority, not only can developing countries guarantee sustainability, but they can also block destructive grabs.

In addition to the base in nature that it needs, sustainable agriculture requires a social and economic base. Current social and economic conditions in Ukraine give mountaineers little to build on, and the impoverished country provides little security or stability for building. The country's kleptocrats threaten to undermine all existing groundwork by plundering both the natural and economic bases of sustainable production. Moreover, mountaineers cannot expect a rescue from abroad. Wealthy Western countries, when not suffering from NGO-fever or pushing supranational governing bodies, shun local knowledge and spout the empty slogans of development. At present, the best and only hopes for mountaineers are their own unnoticed and nimble enterprises, which have survived other periods of economic and political turmoil.

Sustainability in the Carpathians means embracing new knowledge and telling different stories. The typical stories about "cleaning up after communism," which document degradation with shocking statistics and anecdotes, offer simple morals but give no advice on living in the twenty-first century. Insofar as gradual changes, local processes, and natural action did not start or stop with the Fall of the Wall, those interested in sustainability should turn to the landscapes of the Carpathians and other specific places in Central and Eastern Europe. For residents and outsiders alike, understanding places and getting back to place are the first steps in both knowing and living in the contemporary world. Seeking to restore one's place in a Europe that never existed is a guarantee that efforts at sustainability will go nowhere.

Finding the path of sustainability in the Carpathians is difficult. Traveling this route involves surmounting the steep slopes created by past actions and walking on a precipice above the socioeconomic chaos of post-Soviet Ukraine. To chart their route, travelers might benefit from looking at the map laid out in the writings of Aldo Leopold. While it is clear that the conservationist did not know Central European landscapes, he did know sustainability and the value of wildness. For Leopold, sustainable land use was not just a form of use that lasted, but a form that united the natural and cultural "in the joint interest of stability, productivity, and beauty [of the biotic community]."[91] Residents of the Hutsul Region and other bioregions of Central Europe should settle for nothing less.

Notes

1. Several sections of this paper previously appeared in Anthony J. Amato, "In the Wild Mountains: Idiom, Economy, and Ideology among the Hutsuls of Ukraine, 1849–1939" (Ph.D. dissertation, Indiana University, 1998).
2. Aldo Leopold, *A Sand County Almanac with Essays on Conservation from Round River* (New York: Ballantine Books, 1980), 279.
3. On Leopold's Southwest, see Thomas Sheridan, "Human Ecology of the Sonoran Desert," in *A Natural History of the Sonoran Desert* (Tucson: The Sonoran Desert Museum, 2000), 105–6; Dan Flores, *Horizontal Yellow: Nature and History in the Near Southwest* (Albuquerque: University of New Mexico Press, 1999), 29–32.
4. On the links between humans and nature, see Clifford Geertz, *Agricultural Involution: The Process of Ecological Change in Indonesia* (Berkeley: University of California Press, 1963), 10–11; Alice E. Ingerson, "Testing the Nature/Culture Dichotomy in Practice," in *Historical Ecology: Cultural Knowledge and Changing Landscapes,* ed. Carole L. Crumley (Santa Fe: School of American Research Press, 1994), 46–8, 60–3; and Clinton L. Evans, *The War on Weeds in the Prairie West: An Environmental History* (Calgary: University of Calgary Press, 2002), xii–xv.
5. On "captured knowledge," see Dan Flores, *The Natural West: Environmental History in the Great Plains and Rocky Mountains* (Norman: University of Oklahoma Press, 2001), 102–3.
6. Mykola Kulyts'kyi, "Zymarkovi oseli na Halyts'kii Hutsul'shchyny," *Naukovyi zbirnyk Geografichnoï sektsiï pry Ukraïns'kii studenskii hromadi v Krakovi* (Cracow)(1930): 90–2.
7. André de Vincenz, *Traité d'Anthroponymie Houtzoule* (Munich: Wilhelm Fink Verlag, 1970), 31; and Roman Silets'kyi, *Sil's'ke poselennia ta sadyba v Ukraïns'kykh Karpatakh XIX-pochatku XXst.* (Kiev: Naukova Dumka, 1994), 46.
8. *Encyclopedia of Ukraine,* ed. Volodymyr Kubijovyc (Toronto: University of Toronto Press, 1988), s.v. "Hutsul Region."
9. *Skorowidz gmin Rzeczypospolitej Polskiej: Ludność i budynki ... powierzchnia ogólna i użytki rolne. Cześć III Województwo Południowe Serja B, Zeszyt 8-c* of *Statystyka Polski* (Warszawa: Główny Urząd Statystyczny, 1933), 58–9.
10. Volodymyr Shukhevych, *Hutsul'shchyna* (L'viv: Naukove Tovarystvo im. Shevchenka, 1898), vol. 1, 36.
11. See Iakov' Fedorovich Golovatskii [Iakiv Fedorovych Holovats'kyi], *Narodnyia piesni Galitskoi i Ugorskoi Rusi* (Moscow: Moskovskii universitet, 1878), part 1, 586–9; and Volodymyr Kubijovyc, *Volodymyr Kubiiovych, Tom I: Naukovi pratsi* (L'viv: Feniks, 1996), 430–1.
12. See Mar'ian Danylovych Mandybura, *Polonyns'ke hospodarstvo Hutsul'shchyny druhoï polovyny XIX-30-x rokiv XX st.* (Kiev: Naukova Dumka, 1978), 21.
13. *Encyclopedia of Ukraine,* s.v. "Carpathian Mountains"; *Ukraine: A Concise Encyclopedia,* ed. Volodymyr Kubijovyc (Toronto: University of Toronto, 1963), s.v. "Flora"; Oskar Kolberg, *Ruś Karpacka* (Cracow: Drukarnia Uniwersytetu Jagiellońskiego, 1888), vol. 1. Reprint in *Dzieła Wszystkie,* vol. 54 (Wrocław: Polskie Towarzystwo Ludoznawcze, 1963), 180; and Sofron Vytvyts'kyi, *Istorychnyi narys pro hutsuliv,* trans. Mykola Vasyl'chuk (L'viv: Svit, 1993), 31.

14. On transhumance, see Fernand Braudel, *The Mediterranean and the Mediterranean World in the Age of Philip II,* trans. Siân Reynolds (New York: Harper & Row, 1972), vol. 1, 53.
15. For an example, see Tsentral'nyi derzhavnyi istorychnyi arkhiv u L'vovi-TSDIAL (The Central State Historical Archive in L'viv) Fond (F.) 146, Opys (Op.) 64, Sprava (Spr.) 11727, arkush (ark.) 16.
16. Petro I. Arsenych, Iu. H. Hoshko, and Volodymyr Vasyl'ovych Hrabovets'kyi, "Mynule hutsul's'koho kraiu" in *Hutsul'shchyna: istoryko-etnohrafichne doslidzhennia,* ed. H. Iu. Hoshko and R. F. Kyrchiv (Kiev: Naukova Dumka, 1987), 56; and Volodymyr Vasyl'ovych Hrabovets'kyi, *Oleksa Dovbush (1700–1745)* (L'viv: Svit, 1994), 27–30.
17. Mykhailo Lomats'kyi, *Kraïna chariv i krasy. Narysy* (Paris: Ukraïnets', 1959), 213.
18. T. O. Hontar, *Narodne kharchuvannia ukraïntsiv Karpat* (Kiev: Naukova Dumka, 1979), 21.
19. Vytvyts'kyi, *Istorychnyi narys pro hutsuliv,* 25.
20. See "Hutsul's'ka khronika," *Zhovten'* (L'viv) vol. 2 (1964): 157–8.
21. See Vytvyts'kyi, *Istorychnyi narys pro hutsuliv,* 21. In 1902, an estimated 90 percent of all households had to find some source of additional income. See L. O. Olesnevych, "Ekonomichne stanovyshche hirs'koho naselennia Prykarpattia v kintsi XIX i na pochatku XX st," in *Z istoriï zakhidnoukraïns'kykh zemel',* 2nd volume (Kiev: Vydavnytstvo AN URSR, 1957), 145.
22. See Stefan Inglot, ed., *Historia Chłopów Polskich* (Warsaw: Ludowa Spółdzielna Wydawnicza, 1970), vol. 1, 369–70; and Wincenty Pol, *Prace z etnografii Północnych stoków Karpat* (Wrocław: Polskie Towarzystwo Ludoznawcze, 1966), 153.
23. On other areas of Europe, see Jerome Blum, *The End of the Old Order in Rural Europe* (Princeton: Princeton University Press, 1978), 275–6. In the nineteenth century, the potato promoted expansion in the Alps, too; see Robert Netting, *Balancing on an Alp: Ecological Change and Continuity in a Swiss Community* (Cambridge: Cambridge University Press, 1981), 161–3; and John W. Cole and Eric R. Wolf, *The Hidden Frontier: Ecology and Ethnicity in an Alpine Valley* (New York: Academic Press, 1974), 149–51.
24. Pol, *Prace z etnografii,* 152; Hontar, *Narodne kharchuvannia,* 20; Mykhailo Zubryts'kyi, "'Tisni poky.' Prychynky do istoryï Halychyny 1848—1861 rr.," *Zapysky Naukovoho Tovarystva im. Shevchenka* (L'viv), vol. 26 (1898): 1; and John Komlos, *The Habsburg Monarchy as a Customs Union: Economic Development in Austria-Hungary in the Nineteenth Century* (Princeton: Princeton University Press, 1982), 86.
25. Leopold Waigiel, *O Hucułach: zarys etnograficzny* (Cracow: Towarzystwo Tatrzańskie, 1887), 21; Volodymyr Hnatiuk, "Prychynky do piznania Hutsul'shchyny," *Zapysky Naukovoho Tovarystva im. Shevchenka* (L'viv), vol. 123–4 (1917), 37; M. Kossak, "Hutsuly: shkyts' etnografichnyi," *L'vovianyn: pryruchnyi y hospodarskii misiatsoslov na rôk zvychainyi 1861* (L'viv, 1860), 55; Golovatskii, *Narodnyia piesni,* 694; and Pol, *Prace z etnografii,* 153.
26. Fertilizing fields with ashes from brush was still practiced in the 1930s. See Jan Falkowski, *Zachodnie pogranicze Huculszczyzny dolinami Prutu, Bystrzycy Nadwórnianskiej, Bystrzycy Sołotwinskiej i Łomnicy* (Lwów: Nakład Towarzystwa Ludoznawczego, 1937), 82; and Stepan P. Pavliuk, *Narodna ahrotekhnika ukraïntsiv Karpat druhoï polovyni XIX-pochatku st.* (Kiev: Naukova Dumka, 1986), 32–3.
27. Pavliuk, *Narodna ahrotekhnika,* 46.
28. Ibid., 45–7.
29. Shukhevych, *Hutsul'shchyna,* vol. 2, 164.
30. Golovatskii, *Narodnyia piesni,* 694; Mar'ian Danylovych Mandybura and Stepan P. Pavliuk, "Zemlerobstvo," in *Hutsul'shchyna: istoryko-etnohrafichne doslidzhennia,* ed. H. Iu. Hoshko and R. F. Kyrchiv (Kiev: Naukova Dumka, 1987), 109–12; Adam Fischer, *Rusini: zarys etnografii Rusi* (Lwów: Zakład narodowy im. Ossolińskich, 1928), 126; Pavliuk, *Narodna ahrotekhnika,* 72; and Falkowski, *Zachodnie pogranicze,* 31, 81.
31. Mandybura, "Zemlerobstvo," 109; and Falkowski, *Zachodnie pogranicze,* 82.
32. Ivan Drozhdzh, "Hirski pasovyska, tsarynky i polonyny," in *Kaliendar' hutsulskyi na rik 1935* (Warsaw: Tovarystvo pryiateliv Hutsulshiny, 1934), 109; Mandybura, "Zemlerobstvo," 113; and Shukhevych, *Hutsul'shchyna,* vol. 2, 168.

33. See chapter three of Amato, "In the Wild Mountains."
34. Izydor Kopernicki, "O Góralach ruskich w Galicyi. Zarys etnograficzny według spostrzeżeń w podroży, odbitej w koncu lata 1888r.," *Zbiór wiadomości do antropologii krajowej* (Cracow), vol. 13 (1889): 27.
35. See Volodymyr Vasyl'ovych Hrabovets'kyi, *Hutsul'shchyna XIII-XIX stolit': istorychnyi narys* (L'viv: Vyshcha Shkola, 1982), 115; and *Gemeindelexicon im Reichsrate vertretenen Königreiche und Länder. Bearbeitet auf Grund Ergebnisse der Volkszählung vom 31. Dezember 1900* (Vienna: K. K. Hof- und Staatsdruckerei, 1907), 298.
36. *Skorowidz gmin Rzeczypospolitej Polskiej,* 58; Tadeusz Pilat, ed., *Podrzecznik statystyki Galicyi* (Lwów: Krajowe Biuro Statystyczne, 1903), 140–3; and Ivano-Frankivs'kyi oblasnyi derzhavnyi arkhiv-IFODA (The Ivano-Frankivs'k District State Archive) F. 2, Op. 14, Spr. 51, ark. 44, 45.
37. See *Skorowidz gmin Rzeczypospoltiej Polskiej,* 58–9.
38. See *Encyclopedia of Ukraine*, s.v. "Hutsul Region," 286.
39. Roman Włodzimierz Harasymczuk and Wilhelm Tabor, *Etnografia Połonin Huculskich* (Lwów: Towarzystwo Ludoznawcze, 1938), 14; Ie. I. Siavavko, "Tradytsiini narodni znannia," in *Hutsul'shchyna: istoryko-etnohrafichne doslidzhennia*, ed. H. Iu. Hoshko and R. F. Kyrchiv (Kiev: Naukova Dumka, 1987), 261; and Ivan Senkiv, *Die Hirtenkultur die Huzulen. Eine volkskundliche Studie* (Marburg/Lahn: J. G. Herder Institut, 1981), 146.
40. Harasymczuk, *Etnografia Połonin Huculskich,* 5, 45.
41. Shukhevych, *Hutsul'shchyna,* vol. 2, 184.
42. Mandybura, *Polonyns'ke hospodarstvo,* 22.
43. Harasymczuk, *Etnografia Połonin Huculskich,* 13–14, 69.
44. Railroads created links between nature and markets similar to those described in William Cronon's *Nature's Metropolis: Chicago and the Great West* (New York: W. W. Norton and Company, 1991), xv, 92–3. For details on the expansion of railways in Eastern Galicia, see Richard Rudolph, "The East European Peasant Household and the Beginnings of Industry: East Galicia, 1786–1914," *Ukrainian Economic History,* ed. I. S. Koropeckyj (Cambridge, MA: Harvard University Press, 1991), 355; and Stella Hryniuk, *Peasants with Promise: Ukrainians in Southeastern Galicia 1880–1900* (Edmonton: Canadian Institute of Ukrainian Studies, 1991), 47–55.
45. See André de Vincenz, *Traité d'Anthroponymie Houtzoule,* 37.
46. M. S. Hlushko, "Lisozahotivli," in *Hutsul'shchyna: istoryko-etnohrafichne doslidzhennia,* ed. H. Ie. Hoshko and R. F. Kyrchiv (Kiev: Naukova Dumka, 1987), 119, 124.
47. Mykhailo Holubets', ed., *Antropohenni zminy bioheotsenotychnoho pokryvu v Karpats'komu rehioni* (Kiev: Naukova Dumka, 1994), 69–70.
48. *Encyclopedia of Ukraine,* s.v. "Carpathian Mountains," 370; and *Ukraine: A Concise Encyclopedia,* s.v. "Flora," 132–3.
49. Alan Hopkins, ed., *Grass: Its Production and Utilization* (London: Blackwell, 2000), 309.
50. For a discussion of the effects of fertilizers on alpine meadows with edelweiss, see "Grow & Bloom No More?" *Bloomington Voice* (22–29 May 1997): 6.
51. Ihor Kozak and Maciej Augustyn, "The Trends of Anthropogenic Pressure in [the] Polish and Ukrainian Carpathians," in *Selected Ecological Problems of [the] Polish-Ukrainian Carpathians,* ed. Kajetan Perzanowski and Maciej Augustyn (Bieszczady, Poland: Polish Academy of Sciences, 1997), 21.
52. Ihor Kozak and Vladimir Menshutkin, "An Investigation of a Mixed Beech Forest Dynamics in [the] Ukrainian Carpathians Using a Computer Model," in *Selected Ecological Problems,* 23.
53. Holubets', *Antropohenni zminy,* 40.
54. See Stanisław de Vincenz, *On the High Uplands: Sagas, Songs, Tales, and Legends of the Carpathians,* trans. H. C. Stevens (London: Hutchinson, 1955), 170–1.
55. Pavliuk, *Narodna ahrotekhnika,* 46.

56. Holubets', *Antropohenni zminy,* 19, and Philip R. Pryde, *Environmental Management in the Soviet Union* (Cambridge: Cambridge University Press, 1992), 133.
57. Kozak and Augustyn, "The Trends of Anthropogenic Pressure," 21.
58. Holubets', *Antropohenni zminy,* 40, 94.
59. Ibid., 39-40.
60. Ihor Kozak and Mykhailo Holubets', "Lisovyi bioheotsenotychnyi kompleks verkiv'ia Pruta," in *Antropohenni zminy bioheotsenotychnoho pokryvu v Karpats'komu rehioni,* ed. Mykhailo Holubets' (Kiev: Naukova Dumka, 1994), 40.
61. See *Bat'kôvshchyna* (L'viv) vol. 15, no. 13 (1[13] July 1893), 102–3 (newspaper article; no author or title); and Iakov' Fedorovich Golovatskii, "Podorozh po Halyts'kii ta Uhors'kii Rusi opysana v lystakh do pryiatelia u L.," in *Podorozhi v Ukraïns'ki Karpaty,* ed. and trans. Mariia Val'o (L'viv: Kameniar, 1993), 50.
62. See S. M. Stoiko, "Katastrofni ekolohichni protsesy v Ukraïns'kykh Karpatakh i ïkh poperedzhennia," in *Problemy Hutsul'shchyny: tezy dopovidei,* ed. H. Hutsuliak et al. (Chernvitsi, 1993), 96; and Ihor Stebelsky, "Ukraine," in *Environmental Resources and Constraints in the Former Soviet Republics,* ed. Philip R. Pryde (Boulder: Westview Press, 1995), 151.
63. Golovatskii, "Podorozh po Halyts'kii ta Uhors'kii Rusi," 60–1.
64. Holubets', *Antropohenni zminy,* 89.
65. See the description of erosion in Hrabovets'kyi, *Hutsul'shchyna XIII–XIX stolit',* 115.
66. Janusz Gudowski, "Market Oriented Transformation of the Rural Society in Mountain Zones: Example of Carpathian Villages in Western Ukraine," in *Transforming Rural Sector to the Requirements of Market Economy: Examples from Turkey, Poland and Ukraine,* ed. Gülcan Eraktan and Janusz Gudowski (Warsaw: Dialog, 1997), 158.
67. Ihor Kozak, "Antropohenna transformatsiia roslynnoho pokryvu hirs'koï chastyny baseinu r. Prut," *Ukraïns'kyi botanychnyi zhurnal* 47, no. 2 (1990): 60.
68. Gudowski, "Market Oriented Transformation," 157.
69. Kosmach, 24 July 1994. Notes in the author's possession. See the reference in Gudowski, "Market Oriented Transformation," 153. The situation was similar to the situation in socialist Romania described by Ayşe Gürsan-Salzmann in "Shepherds of Transylvania," *Natural History* (July 1984): 42–52.
70. See chapter three of Amato, "In the Wild Mountains." See also Gudowski, "Market Oriented Transformation," 155.
71. Here the coexistence of forestry agriculture was stressed. See Mykhailo Holubets', ed., *Ukrainskie Karpaty: Priroda* (Kiev: Naukova Dumka, 1988), 197.
72. Kozak, "Antropohenna transformatsiia," 60.
73. Holubets', *Ukrainskie Karpaty,* 177–93. On the background of Soviet conservation reserves and parks, see chapters eight and nine of Pryde, *Environmental Management in the Soviet Union;* and Douglas R. Wiener, "The Changing Face of Soviet Conservation," in *The Ends of the Earth: Perspectives on Modern Environmental History,* ed. Donald Worster (Cambridge: Cambridge University Press, 1988), 253–8, 270–1.
74. The focus was on maximizing income, efficient use, and forests' "ecological usefulness." See Holubets', *Ukrainskie Karpaty,* 98; and see Pryde, *Environmental Management in the Soviet Union,* 116–22, 130–5.
75. V. V. Shepa, "Problemy ahropromyslovoho kompleksu hirs'koho rehionu Karpat," in *Tezy dopovidei Mizhnarodnoï naukovo-praktychnoï konferentsiï "Problemy ahropromyslovoho kompleksu hirs'koho rehionu Karpat,"* ed. I. Iu. Fogel' et al. (Velyka Bakta: Zakarpats'kyi instytut ahropromyslovoho vyrobnytstva, 1994), 3–4.
76. V. I. Komendar, *Likars'ki roslyny Karpat* (Uzhhorod: "Karpaty," 1971), 8–9.
77. Ulrich Göttke-Krogmann, "Huzulen—Vergangenheit und Gegenwart," *Österreichische Osthefte* 42, no. 3–4(2000): 124.
78. Gudowski, "Market Oriented Transformation," 153–8.
79. Kosmach, 24 July 1994. Notes in the author's possession.

80. Gudowski, "Market Oriented Transformation," 154.
81. Kosmach, 24 July 1994. Notes in the author's possession.
82. Kozak and Augustyn, "The Trends of Anthropogenic Pressure," 21.
83. S. Kopchak, "Demohrafichni aspekty vyvchennia Hutsul'shchyny," in *Hutsul'shchyna: perspektyvy ïï sotsial'no-ekonomichnoho i dukhovnoho rozvytku v nezalezhnii Ukraïni* (Ivano-Frankivs'k, 1994), 146–7.
84. One scholar has suggested the possibility of the coexistence of tourism and agriculture in the Polish Carpathians. See Włodzimierz Kurek, "Agriculture Versus Tourism in Rural Areas of the Polish Carpathians," *GeoJournal* 38, no. 2 (1996): 191–6.
85. Emily W. B. Russell, *People and the Land Through Time: Linking Ecology and History* (New Haven: Yale University Press, 1997), 18.
86. On swidden cultivation's sustainability, see Timo Myllyntaus, Minna Hares, and Jan Kunnas, "Sustainability in Danger? Slash-and-Burn Cultivation in Nineteenth-Century Finland and Twentieth-Century Southeast Asia," *Environmental History* 7, no. 2 (2002): 267–8; and Stephen J. Pyne, *World Fire: The Culture of Fire on Earth* (New York: Henry Holt and Company, 1995), 86–94.
87. The Carpathians contain both the converted landscapes celebrated by René Dubos, and the wilderness that Edward Abbey argued Dubos failed to appreciate. See Edward Abbey, *Down the River* (New York: E.P. Dutton, 1982), 119–20; and René Dubos, *The Wooing of the Earth* (New York: Charles Scribner's Sons, 1980), 79–84.
88. For an example, see Bjorn Lomborg, "The Environmentalists Are Wrong," *NYTimes.com* August 26, 2002 Opinion.
89. See D. J. Peterson, *Troubled Lands: The Legacy of Soviet Environmental Destruction* (Boulder: Westview Press, 1993), 250–1.
90. On the lack of costs and ownership under socialism, see the comments in Murray Feshbach and Alfred Friendly, *Ecocide in the USSR,* with foreword by Lester Brown (New York: Basic Books, 1991), xii–xiii, 49–50; Jacek Wódz and Kazimierza Wódz, "Environmental Sociology in Poland and the Ecological Consciousness of the Polish People," in *Environment and Society in Eastern Europe,* ed. Andrew Tickle and Ian Welsh (New York: Longman, 1998), 101; Philip R. Pryde, *Conservation in the Soviet Union* (Cambridge: Cambridge University Press, 1972), 42–4; and Peterson, *Troubled Lands,* 16.
91. Leopold, *Sand County Almanac,* 199.

Chapter 13

NEW APPROACHES TO SUSTAINABLE COMMUNITY DEVELOPMENT IN RURAL SLOVAKIA

Slavomíra Mačáková

Introduction

Slovakia combines features of both a developed nation and a developing country. Taking into account its economic structure, which is dominated by industry and services and boasts a relatively high level of GDP per capita, and its quite well developed social security and social safety net systems, the country is comparable with other European countries. On the other hand, marginalization of some rural and minority communities, cases of extreme poverty, and relatively frequent occurrence of poverty-related diseases, are typical signs of a developing country.

Since the collapse of the socialist regime in 1989 and, more visibly, since the split of Czechoslovakia in 1993, Slovakia has been undergoing a comprehensive transition process from a centrally planned command system to a market-based democratic society. Unfortunately, as has become apparent over the last decade, Slovakia has not been entirely successful in addressing the various challenges inherited from the previous regime or brought about by this transition. The policies that have facilitated the transformation processes have not always been coherent and decisive. They have come at high social and economic costs. As a result, the country exhibits significant challenges in various aspects of sustainable development and in respect to large regional differences in economic and human development, which in some regions are even further exacerbated by the appalling social and economic situation of the Romany minority.

Roma populations have long been amongst the poorest, most marginalized and vulnerable in Slovakia. While some progress was made in narrowing the gap between Roma and non-Roma populations during the socialist period, the economic conditions of Roma have more often declined in both relative and absolute terms during the last fifteen years. This situation is reflected by insufficient housing, high unemployment rates, deteriorated health, low rates of educational attainment, and low levels of political participation. These negative trends in social and economic conditions of the Roma combine with cultural barriers, such as language, to affect the access of Roma families to social services, education, and employment.

The Scope of the Problem and Challenges to Sustainability

Demographic Data

Despite the fact that data on Roma are rather unreliable throughout the country, it can be stated that Slovakia has one of the largest Roma populations in Europe. Although in the 2001 census only 89,920 individuals—1.7 percent of the total population of Slovakia—declared themselves to be Roma, informal estimates suggest that there are approximately 320,000 Roma in Slovakia, thus constituting about 6 percent of the population, the second-largest minority in Slovakia.[1]

Slovakia is divided into eight self-governing regions that differ greatly in the majority of indicators for economic and human development. Significant regional disparities and notable variations between ethnic groups are observed in the unemployment rate. By the end of 2004, the national unemployment rate was 13.1 percent. While unemployment in the Bratislava region was less than 4 percent, it was over 20 percent in the Eastern Slovakian regions of Košice and Prešov (rising as high as 50 percent in certain parts of these regions).

Even though official unemployment data by ethnicity are lacking, a 1997 survey prepared by the Ministry of Labor, Social Affairs and Family estimated that Roma comprised between 17 to 18 percent of the total unemployed in 1996. However, in the eastern region unemployment was as high as 40 to 42 percent, reaching 100 percent in almost all segregated Roma communities.[2]

Human Development Index

According to the HDI (Human Development Index), Slovakia belongs among the group of countries with high human development, ranking fortieth in the world. However, if HDIs for each region and different ethnic group were available and taken into account, the numbers would be quite different.

While absolute poverty in Slovakia has been found to be low, there are significant pockets of poverty, concentrated especially in regions with large Roma populations and isolated rural areas. The gap between the greater Slovak population and the "invisible minority" of Roma who live under "survival conditions" is widening, according to the *EU Twinning Program Report*. Although there are fewer people living under the USD 4.30 PPP (purchasing power parity) per person per day in Slovakia than in neighboring Poland and Hungary, there are twice as many people who live in severe poverty, i.e., under USD 2.15 PPP per person per day.[3]

According to the World Bank report on *Poverty and Welfare of Roma in the Slovak Republic*,[4] the poverty among the Roma is closely linked to four main factors: (1) regional economic conditions; (2) the size and concentration of the Roma population in a settlement; (3) the share of Roma in a settlement; and (4) the degree of geographic integration or segregation of the settlement and its proximity to a neighboring village or town.

Roma Settlements

Roma live in a range of different types of settlements, whose characteristics depend on the degree of their segregation/integration (both spatial and social) and the geographic location of the settlement (rural or urban environment). An estimated one fourth of Roma in Slovakia live in a settlement called an *osada*—usually a socially and economically downtrodden cottage colony, located on the outskirts of a village or town. Based on the information from the sociographic mapping of Roma settlements in Slovakia prepared by the Office of Plenipotentiary for Roma communities, a total number of 1,575 *osada*s were identified across the whole of Slovakia, perceived by the majority population as the settlements of Roma communities. In 772 cities or towns Roma live integrally with the majority population. According to the location, the remaining Roma communities can be devided into:

- Settlements in the towns and cities(168),
- Settlements situated at the edge of a town or city (338),
- Settlements located at some distance from a town or city or separated by a natural or artificial barrier (281).

Of the latter, 149 segregated *osada*s—those located at some distance from a village or a town—are without a source of drinking water. Overall, 81 percent of Roma *osada*s do not possess a sewage or sanitation system, gas heating is unavailable for 59 percent, and 37 percent lack a source of drinking water. The majority of the *osada*s are concentrated in the eastern part of Slovakia. Dwellers in the *osada*s are hampered in gaining basic services (water, sanitation, heat and energy) mainly by poverty and by the absence of interactions with local authorities.

Characteristics of the Roma Population

The Roma population in the *osada*s is significantly younger and has been growing more rapidly than other ethnic groups. This is due not only to higher birth rates but also to the fact that many Roma have returned to *osada*s from cities after losing jobs, taking advantage of the availability of cheaper housing. In 1988, 14,988 people lived in *osadas*, but by 2004 this figure had increased to approximately 160,000.[5]

The demographic profile of the Roma is considerably different from that of the total population. The average age amongst Roma is 24.43 years. While in the non-Roma population children under fifteen years make up 17 percent of the inhabitants, in the Roma ethnic group their share is 37 percent. Meanwhile, only 2 percent of Roma are older than sixty-five, compared to more than 12 percent of non-Roma.

Social Situation of the Roma

Stable, permanent employment is almost nonexistent for Roma. Formal employment opportunities for Roma in segregated *osadas* are limited to temporary and seasonal jobs in agriculture, construction, and public works organized by the local authorities. Many Roma have to support their families by working in the informal sector: salvaging, collecting and selling scrap metals, playing music, petty trading, begging in larger cities, or working without job permits for employers who want to avoid paying taxes and state insurance contributions. Many Roma in Slovakia complain that discrimination is a significant barrier to obtaining employment and a rationale for not searching for work outside of their communities. Moreover, many young people and also their parents do not see any benefit in obtaining high school education in a situation where the young people remain unemployed after graduation. Social assistance benefits are thus becoming the only source of income for the majority of Roma individuals and Roma families. In addition to that, the social security system in Slovakia provides people with little incentive to actively participate in the labor market and disinclines them to seek jobs or initiate income-generating activities in order to improve their life conditions.

Local Authorities

Many local officials in the economically deprived towns and villages throughout the country are unaccustomed to communicating and collaborating with Roma. Negative stereotypes, on both sides, can lead to limited contacts and interactions between Roma and non-Roma and to a near-complete absence of relations between the Roma and local government officials. While most of the *osadas* often are without water, gas, or electricity, many mayors have had little experience interacting with individuals from these settlements, let alone forming partnerships with them to improve the situation.

International Perspectives

The problems presented above are widely recognized by national and international stakeholders interested in securing sustainability for social and economic development in Slovakia. Slovakia's membership in the European Union makes it even more urgent to address problems of regional disparities and the social disintegration of the Roma. The lack of equal services for Roma and their insufficient political representation could impede conformance to the minority rights standards of the EU. This is why the overriding goal of the Slovak government in the last few years has been to search out the best ways to mitigate the social and economic deprivation in certain regions, eliminate interethnic tensions, promote the development of a multicultural society, and change the attitude of the mainstream population towards the Roma.

Not only has the EU been strictly observing the enforcement of human rights and respect for minorities, but it is also providing assistance to Slovakia, mostly through its EC Program funding opportunities. The EU cofinanced various projects involving Roma under the PHARE support program, and is particularly active in the areas where the Slovak national government has failed to fulfill its commitments. Other donors have also been active in providing assistance to Slovakia in order to deal with this crucially important issue; they include the Department For International Development (DFID, UK), Canadian International Development Agency (CIDA), U.S. Agency for International Development (USAID), Open Society Institute (OSI), United Nations Development Programme (UNDP), MATRA (Netherlands), and others.

The Case of the Middle Spiš Region

One of the regions in Slovakia that has had the most difficulty with the transformation process and economic restructuring is the Middle Spiš region in Eastern Slovakia, encompassing the Gelnica, Levoča, and Spišská Nová Ves districts. Consisting of 140,000 inhabitants (16 percent of them Roma), the Middle Spiš is the best example of regional socio-economic retardation and deprivation in Slovakia. In the absence of coherent regional and local development strategies, business entities have had little success restructuring themselves and attracting domestic or foreign investments. The national, regional, and local authorities have failed to create a supportive environment in which entrepreneurs would benefit, flourish, and create job opportunities for the unemployed.

Summary

In addressing the challenges and deficiencies outlined above, there is an urgent need to elaborate a comprehensive, coordinated, and holistic approach that can deal effectively with regional and social disparities in development within Slovakia, particularly in regard to rural areas and the Roma population. The new approach needs to ensure that all segments of the society will be able to benefit from opportunities brought about by the transition process, economic development, and integration with the EU. To achieve this, both domestic efforts and international assistance needs to be mobilized.

The YOUR SPIŠ Program: Basic Information

In order to address this situation, the United Nations Development Program (UNDP) has initiated the local development program *YOUR SPIŠ—Sustainable Community Development in the Middle Spiš*. The program seeks to contribute to local development and social integration through building capacities and networks at the local and regional level. It is implemented by the local nonprofit organization ETP Slovakia in cooperation with the Office of Government of the Slovak Republic, with financial and institutional assistance from the UNDP and the government of the Netherlands through its MATRA program.[6]

Since its start in January 2001, the program has created a unique opportunity for governmental officials, mayors, community groups, public and private sector leaders, and other actors from the region to develop and test new comprehensive approaches to dealing with the complex and challenging issues of social and economic development.

Goals and Objectives

The goals of the YOUR SPIŠ program are to strengthen the technical, economic, and human resources in the Middle Spiš region through the establishment of a sustainable local social and economic infrastructure, to support the integration of disadvantaged groups and ethnic minorities into mainstream society, and to involve them in viable economic activities. Also, the program assists the region by providing the seed capital necessary to start up these initiatives.

The YOUR SPIŠ Program has four main objectives or specific approaches to be used in order to reach the stated goals:

1. Enhance social and human capital in the Middle Spiš by building up the capacities of disadvantaged groups and fostering the development of community-based organizations as the basis for local social and economic initiatives to improve livelihoods;
2. Establish Community Development Centers in close cooperation with local governments, which will foster community partnerships, exchange of local and community knowledge, and development of income-generating activities;
3. Establish a Community Development Facility as a training, financing, and marketing mechanism that will provide seed capital to small entrepreneurial initiatives and micro-grants for those activities that benefit the whole community; and
4. Utilize the experience of YOUR SPIŠ for the elaboration of national policies on regional development and social integration as well as regional development strategies that will support the inclusion of disadvantaged groups into local economic initiatives.

The Critical Role of Partnership

No programs like YOUR SPIŠ have been implemented previously in Slovakia; as such, this is a pilot program. It builds upon the national, regional and local governmental bodies' commitment to working out and promoting a results-oriented, sustainable approach to regional and local economic development that is firmly based on partnership among government, private sector, and civil society organizations. In order to achieve its goals and objectives, the program takes a people-centered, institutional and integrated approach rooted in a process of social mobilization that is inclusive of all ethnic groups.

Experience in the Field

The designers of the program used experience from projects implemented in Asian countries as a foundation and attempted to transform this experience to suit Slovak conditions. From the earliest stages of program implementation, the YOUR SPIŠ management team encountered an immediate challenge—to make effective use of the experience and knowledge of the local people when adapting the program to local conditions, i.e., tailoring it for optimal realization of the program goals in each place. This required a lot of flexibility and a participatory approach. This approach, always approved by all program partners, has been confirmed as the best way to generate benefits for all stakeholders and interested partners in the Middle Spiš region.

The Approach

YOUR SPIŠ builds on social mobilization as a means to foster local development. Social mobilization is a tool to motivate the disadvantaged groups and the Roma in the program's thirteen partner municipalities (Iliašovce, Levoča, Mníšek nad Hnilcom, Nálepkovo, Prakovce, Rudňany, Spišská Nová Ves, Spišské Podhradie, Spišské Toamšovce, Spišský Hrhov, Spišský Štvrtok, Žehra, and Švedlár) in the three districts of the Middle Spiš (Gelnica, Levoča, and Spišská Nová Ves). The aim is for target groups to improve their livelihood by utilizing local resources.

The Target Groups

YOUR SPIŠ targets marginalized/disadvantaged groups in the region. These are, for instance, families and individuals whose per capita income is below the minimum income standard and whose social situation and living conditions (housing, level of education, ability to provide and care for their family) are below the average level of the society. Young families and other individuals who are long-term unemployed and whose only source of income is social welfare or social benefits, constitute a large part of the target beneficiaries as well. YOUR SPIŠ pays special attention to Romany people, almost all of whom are chronically unemployed and totally dependent on the social welfare system. Many of these people are currently unable to change their economic situation through their own abilities and resources.

Implementation

Community activists employed by the YOUR SPIŠ program work together with the members of disadvantaged groups from selected municipalities. Upon entering individual towns and villages, they initially inform residents about the objectives of the YOUR SPIŠ program. Then they involve the local authorities in the program's implementation, in order to gain the support of local leaders and state institutions. Most difficult, perhaps, is to earn the trust of the target group. The community activists dedicate significant time to visits and discussions with community members, with the aim of mapping selected localities and assisting in improvement of the situation in them, including infrastructure, social services, natural services, economic power, and human resources. The activists also identify the needs, interests, goals, and abilities of the target beneficiaries. Groups and individual community members are asked to share their views on the challenges and problems affecting their personal lives and the community as a whole. They are also asked to identify opportunities for improving the situation.

The discussions with a wide variety of people soon engender, usually, a good awareness and understanding of the program goals and objectives among the general public and key people within the respective municipalities. It has been revealed that the top-priority challenges to be tackled within the poor and disadvantaged communities are the improvement of housing, employment, education, and health conditions for their members.

The leading principle is that when the local people themselves are motivated to attempt to improve their situation, the program management team will provide expertise and technical assistance. Men and women in the village, settlement or neighborhood, covering both Romany and non-Romany, are empowered to organize themselves into self-governing, independent community organizations through a process of social mobilization. The purpose of establishing such organizations is to strengthen the principles of self-help and cooperation, to bring about a general improvement in the quality of life, to reduce the deepening of social difference, and to create solidarity, trust and cohesion amongst the population. It is a demanding and long-term task. It entails overcoming nihilism, apathy, lethargy, and people's lack of confidence in their own abilities, in identifying and empowering their own resources, and in discovering internal strengths such as manual skills, artistic talent, managerial skills, and entrepreneurial abilities.

The community activists always take into consideration the level of maturity and readiness of the members of the groups, so that a positive change will occur as a result of the community members' own will and out of genuine interest. Consequently, the target beneficiaries establish community-based organizations, associations, unions, or other types of formal and informal groups as a means of participation in and fostering of local development. The program staff also supports pre-existing groups and organizations to be more actively involved in community development.

The formation of community organizations encourages and facilitates the genuine participation of both men and women in the process of identification and generation of their own resources and utilization of their own potentials. The community members make decisions to resolve common development problems and jointly plan, develop, and undertake social and economic development initiatives. The social mobilization process also results in the mobilization of psychological strengths (e.g., self-confidence) among the members of the communities.

Members of the community-based organizations are then supported in developing organizational rules and management structures for their organization. The YOUR SPIŠ program ensures training in leadership and management for community members and at the same time provides the technical support to guarantee successful implementation of the community-based initiatives designed by the program's target beneficiaries in cooperation with the local authorities and other local stakeholders.

The existing and newly established community organizations from the municipalities involved in the program have the opportunity to apply for grants for planned activities leading to the development and enhancement of public life in the community. Individuals interested in conducting small-scale or start-up entrepreneurial activities have the opportunity to receive financial support in the form of micro-credits from the program.

Cooperation with Local Government

All of the activities of the program are regularly negotiated with local government representatives. Based on the latter's suggestions, program activities are adjusted so as to best fit local needs. This approach has produced concrete results within a relatively short period. Less than two years after becoming involved in the program, the same mayors, who previously were not accustomed to communication and cooperation on developmental projects with the poor communities, have already agreed to partner with several Romany organizations from their villages or towns.

This is thanks largely to the three-tiered approach to building partnership, which brings together nonprofit organizations, local authorities, community groups, and the private sector. All community projects that ask for financial contributions from the program are negotiated with local authorities beforehand, and the authorities provide only 50 percent of the support delivered to individual projects, in the form of material or financial resources. This way, the recipient community has to mobilize its own resources and maintain an investment in the success of a particular project. On the governmental side, the mayors are generally cooperative and supportive. Excellent relations with the local, district, and regional government representatives are a driving force behind the active work of individuals involved in the program.

Summary of Program Outcomes: Some Basic Facts

The four and a half–year-long YOUR SPIŠ program has almost reached the end of its operative time span. Thus far in the partner villages and towns, it has helped to form more than fifty informal community-based groups, about twenty of which have already transformed themselves into formal organizations—registered civic associations. Thanks to the endeavors of the YOUR SPIŠ management team, the attempts to reach a consensus between the majority and the Roma minority are becoming more and more successful. These have resulted in small joint projects or even in the creation of ethnically mixed community organizations in selected municipalities.

Furthermore, YOUR SPIŠ has mobilized funding support from the government public works program, allowing some long-term unemployed individuals to be paid to work for the benefit of YOUR SPIŠ program activities. As such, YOUR SPIŠ also contributes to enhancing the skills and opportunities of the long-term unemployed with a view to their reintegration into the labor market. One example of this support is that in each village or town, the mayor has employed at least one person through the newly established Social Development Fund (European Social Fund)—an assistant to community activists who helps prepare, organize, and conduct meetings with the members of the target beneficiaries in the given village or town. Moreover, in all of the program municipalities a total of more than five hundred additional people have been and are being funded through the public works program to work on small community projects associated with YOUR SPIŠ.

Thanks to the strong partnerships and new cooperation instigated by the program, more than 150 small community projects have been initiated and implemented in YOUR SPIŠ towns and villages. The small projects are being funded jointly by the UNDP, MATRA, local authorities, and the members of the community. The aim of the grant assistance is to support community organizations and initiatives by community members that lead to an enhanced quality of life in the municipalities involved in the program. Projects submitted to the grant round should comply with municipal development plans and should involve as many community members, including the Roma, as possible, through voluntary work. Projects should benefit the entire community and support a wide spectrum of civic initiatives in the program municipalities. From the YOUR SPIŠ grant scheme, USD 200,000 has already been invested; a total of work valued at over USD 300,000 either has been or is going to be realized.

The Main Categories of Project Activities

Smaller projects have been undertaken within several topical fields or categories:

1. Enhancement of the municipality: cleaning of public areas, planting greenery, waste disposal, landscaping, bridge construction, forest park establishment;

2. Education of members of community organizations: developing social dialogues, building communication skills, introducing computing, basic hygiene, and fundamental business skills;
3. Renewal and maintenance of folk traditions: developing traditional handicrafts, weaving workshop, crafts, smithing, pottery;
4. Development of cultural life in the municipalities: preservation of Roma culture, participation in theatre ensembles, continuation of female folk dress;
5. Completion of sports facilities and support of sports activities in municipalities: football, skating rinks, fitness.

Those who decide to transform their ideas into projects but do not have sufficient experience with project development and writing have the opportunity to receive training, assistance, and consultation from the program.

Selected Examples of Projects

These community projects serve as examples of people in the program partnerships developing the will to cooperate in a number of areas that bring together Roma and non-Roma and help alleviate the tensions between the two groups.

In Spišský Štvrtok, local people built a small bridge that joined the Romany and non-Romany sections of the village. Also, with the financial assistance of YOUR SPIŠ, Romany and non-Romany children from the village practice and perform cultural programs for mixed audiences of adult Roma and non-Roma and so foster positive relationships between the majority and the minority.

Another group, The Roma Way from the village of Iliašovce, was dedicated to the repair and redesign of Roma housing units, construction of a water supply that would provide drinking water for Roma households, and improvement of the landscape and local environment. Roma Iliašovce, a new Romany civic association, focuses its activities on landscape improvements in the settlement and agriculture activities such as livestock breeding, with the ultimate goal of creating self-employment job opportunities.

In Levoča, a newly created Romany civic association cooperated with the local authority on a "Sidewalks and Landscaping" project, which dealt with the repair of sidewalks, mowing grass, cleaning grounds, landscaping, and improving the common areas in front of city houses. Through these voluntary project activities, the group mobilizes and activates the residents of the Romany neighborhood, developing their work skills and involving them in the search for solutions to community problems.

Another project, in Nálepkovo, helps Roma women gain enough practical skills and experience in traditional fabric weaving to allow them to generate needed income for their families and thus become less dependent on social welfare.

Several ideas for sustainable agriculture and forestry are being developed and implemented in close cooperation with mayors and local leaders. Not only do these ideas bring about positive attitudes in taking care of the environment, but they also create sustainable jobs for rural people.

Tackling the Interethnic Tensions Effectively

In tandem with these specific projects, the building of relationships between Roma and non-Roma in itself has begun to mitigate interethnic tensions in the villages and cities. Thus, the level and quality of cooperation between Roma and non-Roma on specific project activities that interest all inhabitants of the village is improving. Furthermore, disadvantaged people involved in the project through volunteer activities or the public works program are assisted in reintegrating into the labor force through involvement in retraining and vocational training courses.

The Micro-Credit Program

Another financial activity developed within the framework of YOUR SPIŠ is a micro-credit program, which assists in reducing the high unemployment rate in the Middle Spiš. The micro-credits are small loans designed to support small self-help and start-up entrepreneurial initiatives, as well as activities that may generate income for the members of the community.

Members of community organizations and other individuals who take advantage of this opportunity can, through the micro-loan program's training courses, gain the knowledge necessary to start their own entrepreneurial activities. These courses are both theoretical and practical in nature, and provide the participants with necessary basic skills. Having completed this training, participants are eligible for financial support from the micro-credit scheme and may start up their own small, income-generating activities.

The Sow Breeding Project

Sixty Roma families have been involved in a "gift-for-gift" sow breeding project since its beginning in 2002. Developed in collaboration with Heifer International, an NGO dedicated to the promotion of family-scale animal husbandry enterprises, this experiment was tailored to meet particular needs of the poor Roma and has had a positive impact on the Roma communities. It allows Roma to feed themselves, secure financial income for their families through their own effort and work, and become self-reliant.

Evaluation of the Program

The experience gained during the implementation of the YOUR SPIŠ program has spotlighted several advantages of the social mobilization approach. First, it is people-centered: the focus is the fulfillment of the needs, desires and aspirations of the target beneficiaries. The community activists it cultivates and funds are highly motivated, dedicated people who use their best abilities in their systematic daily fieldwork for the good of the target beneficiaries. Furthermore, community activists initiate various additional activities (social field work, assistance to Roma pupils in local schools or kindergartens, assistance to local physicians with Roma patients, advising the local authority on Roma issues, to name a few) in their respective municipalities in cooperation with local authorities; these too help them succeed with their work in the community. All of these people work together as a team to submit various joint proposals for improvement in the livelihood of their community's Roma people to the local authority, other public authorities, or to external donors.

In addition, the program itself has initiated legislative amendments to social and employment laws, such as tax relief for employers providing jobs to the unemployed, scholarships for talented children from socially disadvantaged families, and numerous other forward-looking measures. Moreover, because the program with its bottom-up approach is managed in a flexible way, its tasks are easily adaptable to the changing situation in the region. The strong, three-tiered partnership is built at the local, regional, and national levels.

Finally, the program has enabled various donors and stakeholders to cooperate on various components or specific projects within the YOUR SPIŠ program, so that the program as a whole has been able to provide a model for holistic, integrated intervention in this region. Knowledge of the best examples of successful Roma projects and activities is being disseminated throughout Slovakia. Already, these success stories are assisting in alleviating interethnic tensions between Roma and non-Roma.

Conclusion

While Slovakia's poorer communities are plagued by many deep-rooted problems (infrastructure, unemployment, low economic power, poor sanitary conditions, low level of education, and so on), there also are many opportunities that spring from the availability of natural and human resources: considerable skills and experience, a substantial labor force, good land, chances for greatly increased cooperation between the poor and officials from the municipalities, and the marketing potential of local people's products.

The program's results to date show that even small activities prepared and implemented by individuals and small community organizations can foster local development. Through strong partnerships and new approaches, YOUR SPIŠ makes meaningful strides toward decreasing regional and ethnic gaps in sustainable development of the rural areas in the Middle Spiš. It also proves that even the most disadvantaged social groups or areas can mobilize their resources and improve their lives, thus contributing to building sustainability in their communities, regions, and country. This is a very important lesson to share with other disadvantaged communities or regions both within Slovakia and worldwide.

Notes

1. M. Vašečka, "Rómovia," in *Slovensko, Súhrnná správa o stave spoločnosti*, ed. M. Kollár and G. Mesežnikov (Bratislava: Inštitút pre verejné otázky, 2000).
2. I. Radičová et al., *Chudoba Rómov a sociálna starostlivosť o nich v Slovenskej republike*, World Bank Report on *Poverty and Welfare of Roma in the Slovak Republic* (Bratislava: INEKO, Nadácia S.P.A.C.E., Nadácia otvorenej spoločnosti, 2002), 25–8. The English version is available at http://siteresources.worldbank.org/EXTROMA/Resources/povertyinslovak.pdf
3. R. Džambazovič et al., *National Human Development Report—Slovak Republic 2000*, UNDP 2000, available at http://www.cphr.sk/english/undp2000.htm
4. Radičová et al., *Roma Poverty*, 11.
5. Džambazovič et al., *National Human Development Report*.
6. The Netherlands' MATRA program derives its name from the Dutch "maatschappelijke transformatie" or "social transformation." Information on the purposes and operations of the program can be found on the official Netherlands Foreign Ministry site at http://www.minbuza.nl/default.asp?CMS_ITEM=MBZ409267

Chapter 14

SUSTAINABLE DEVELOPMENT IN MORAVIA
An Interpretation of the Role of the Small-Town Sector in Transitional Socioeconomic Evolution

Antonín Vaishar and Bryn Greer-Wootten

Introduction

The various meanings of sustainable development—complicated by its "triple bottom line" of economic, environmental, and social sustainability objectives—take on a particular resonance in the postcommunist Central European states.[1] Whereas economic goals often dominate the discourse, and often in a pessimistic mode, the environmental legacies of the socialist era are quite evident, and the goals of (re)building civil society would appear to confront formidable obstacles in constructing a sustainable future. A salient feature of much of this discourse is the continuing importance placed upon the role and conditions of the state as the unitary object of analysis. It is the contention of this paper that the focus of discussion should be relocated to the regional level, where the objectives of sustainability can be realized to a greater extent by building upon the locality imperatives of sustainable development.

Although "region" was one of the central organizing concepts for the discipline of geography until the early 1960s, it has only recently regained its importance, albeit in a changed form.[2] In part, this relates to the desire to find a context for "locality" in the face of the "globalization" of economic forces as well as one of the assumed consequences of globalization—a diminished role for the nation-state.

In particular, however, the revival of regional concerns can be attributed to the increased attention paid to environmental issues over the last thirty years. The "new regionalism,"[3] then, should incorporate "sustainability" as a broad concept that relates to both economic processes and environmental concerns (*via* its operationalization as "sustainable development"), as well as to social objectives.[4] Importantly, the new regionalism relates primarily to issues of governance,[5] an important linkage to the practical implications of the movement toward sustainability.

In addition to our previous contention that there should be a relocation of discourse to the regional level of concern, we equally assert the necessity of a conceptual framing of the issues under consideration. There are two primary stimuli for this decision: 1) much of the current discourse is phrased directly or indirectly in political terms, and, important as this is, some more fundamental questions are brought into play once one implicates the environment (however defined) in the discussion; and 2) most of the terms employed in the various debates are themselves contested. Space limitations, unfortunately, prevent a full examination of these theoretical issues, but they underpin our conclusions and are subject to further scientific collaboration in the context of the small towns study.

The structure of this chapter is as follows. Following this introduction, the second section outlines briefly some of the problems in using older paradigms (read: disciplinary) to understand recent attempts to integrate environment and economy in sustainability discourse, which demands interdisciplinary perspectives.

The third section is more empirical in nature: a continuing systematic study of the small-town sector (municipalities with town status and populations of less than 15,000) in Moravia, initiated by the Institute of Geonics in 1999, is discussed. The effects of forty years of socialist central planning (1948–1988) are outlined, especially the ways in which the traditional central place structure of settlements was disturbed by implantations of heavy industry and immigrant populations. The principal focus is placed upon changes in this system in the so-called transition period since 1989. Such changes can be related to the effects of globalization, as seen in the development of a market economy, but they are further complicated by the dissolution of the former Czechoslovakia in 1993, as well as the question of accession to the EU in the early years of the new millennium.

The empirical context for discussions of the future of the small-town sector in Moravia (section four) is, then, quite complex in political-economic terms. Factoring in the environment (*qua* sustainable development) only adds to such complexity.[6] In asserting the important role that the small-town sector may play in a sustainable future for Moravia, it is necessary to recognize the crucial relevance of civil society in negotiating the real transformation to sustainability. Fortunately, the Czech history of popular engagement in organizations related to landscape heritage and nature conservation[7] demonstrates locality-based "resistance" to larger-scale powers, including those of the socialist era. In this way, the importance of social capital in a regional setting is highlighted.

Old Paradigms, New Realities

In a recent, seminal overview of the field of economic geography, Scott indicates clearly the extent to which changes in research approaches relate to changing patterns of economic, social, and political development.[8] Thus, many of the traditional studies of settlement patterns, based largely on Central Place Theory, can be viewed in the context of the so-called Fordist patterns of capital accumulation that characterized Western economies in the post–World War II period up to the beginning of the 1970s. The subsequent oil/energy shocks of the early 1970s and the emergence of a "rust belt" of older industrial areas signaled the development of post-Fordist economic patterns. The relatively strong thrust towards locality studies, albeit a minor deviation from dominant political economy approaches, was seen as one result of the new trends.

As indicated earlier, the most recent changes of an economic nature are reflected in various interpretations of the forces of globalization, which in turn have had an interesting dual effect on research paradigms, as noted by Clark, Feldman and Gertler: "For some analysts, difference and the heterogeneity of the economic landscape demand close, detailed analyses of the particular attributes of certain firms, regions, and industries. For others, particularity has to be balanced against larger economic forces operating at higher spatial scales: there is a tension between the local and the global and between fine-grained case studies and stylized facts."[9] The same authors also point to the fact that older paradigms indirectly reflected assumptions concerning the relative stability and predictability of economic forces—acceptable then but no longer tenable in the current situation—and assumed also a relatively neutral status for "location": "Past theories of location have treated the firm, the community, even whole regions as undifferentiated 'black boxes'."[10] One outcome is that most current research programs take the geographical and institutional organization of economic activities on a regional basis seriously into consideration.

Nonetheless, there are inevitably some continuing traditions in terms of research approaches, although there are few empirical studies that directly attempt to disentangle the ways in which globalization, for example, changes expectations that might be derived from more traditional Central Place Theory. One such expectation could be that larger cities and metropolitan areas in nation-states would become increasingly dominant in the national settlement structure. Indeed, this has been demonstrated for the evolution of the urban hierarchy in England and Wales over the period 1913–1988.[11] Similarly, Claval has proposed that recent changes in communications technology have changed the forces influencing urban networks, with increased tendencies toward metropolitanization as a result.[12] Financial services as well have assumed a greater role in recent economic change, and the ways in which urban networks have facilitated (and responded to) such changing conditions have been described by Parr and Budd.[13] Much of this work is based on older paradigms, but it can also be seen as part of the recent "global cities" research program.[14]

Clearly, in any settlement system, changes in the determinants of an urban hierarchy will change the outcome. Recent communications technologies change distance relations for consumers and service providers, consequently influencing patterns of travel behaviors as much as influencing the institutional organization of retailing and other service sectors. Such changes, however, tend to be incremental and relatively slow because there is, in the physical representation of the system, a physical inertia that also affects patterns today. In the Central European countries in the socialist era, there was an added determinant in the form of centralized planning systems that often employed aspects of Central Place Theory in terms of the administration of settlement systems, for example in Hungary.[15] Various examples for the Moravian case are discussed below.

More generally, the role of small towns in regional planning has often been promoted as a platform for the development of marginal areas. One conclusion of a recent comprehensive overview of this research approach (and its subsequent application to planning policy) by Hinderink and Titus is instructive: "The lesson to be learned then is that if we want to acquire insight into the potential development role of small urban centers, *the regional context should play an essential role in our analysis* [emphasis added]."[16]

In fact, regional concerns are evident in much of the European research carried out on small towns, mostly by geographers of German-speaking countries.[17] For the case of Moravia, to which we may now turn, some relevant references of a general nature include Vaishar and Kallabová (2001) and Kallabová (2001), whilst some specific case examples are presented in Vaishar and Zapletalová (1998) and Vaishar et al. (2001).[18]

The Small Town Sector in Moravia: Demographic Context

The recent (2001) population census in the Czech Republic resulted in the following size distribution for settlements (characterized in four groups: see Table 14.1). One notes the more or less even distribution of villages, small towns, larger towns and cities, but the small-town sector (2,000–19,999 persons) contains the largest share (29 percent) of the total population. Quite possibly, this is a typical Central European rank-size distribution, resulting perhaps from planning processes in the socialist era. The inertia represented here reflects the trends at that time toward reducing disparities in living conditions between urban and rural areas: small towns received subsidies to develop original or new local industrial activi-

Table 14.1 Population by Rank-Size of Communities (2001)

Category	Total population	Share [%]
0–1,999	2,672,825	26.0
2,000–19,999	2,983,560	29.0
20,000–99,999	2,489,927	24.1
100,000 and more	2,146,621	20.9

Source: Population and Housing Census March 1, 2001 (Prague: Czech Statistical Office, 2002).

ties, and witnessed the construction of new apartment buildings and the installation of new services.

Such a rank-size distribution is unlike that of most developed Western societies, and the current tendencies toward counter-urbanization have contributed to an increase in the number of people living in small towns. The large cities in the Czech Republic, however, remain the sources of economic and political impulses in the country, largely through their quaternary functions, and a number of people from small towns commute to work there. Still, small towns are important in maintaining a relatively equal population distribution, ensuring at the same time that small villages in marginal areas do not fall into decline. As such, the small-town sector can be described as a dynamic force in the population structure of the Czech Republic.

Small Towns in Moravia: Study Design

The empirical study is based on a sample of small towns in the historical region of Moravia, each with a population of 15,000 or less. In Moravia in 2001, there were 108 such small towns: the majority (fifty-nine) had a population of less than 5,000, and only twelve had populations of more than 10,000. The total population living in these small settlements is relatively stable: 598,995 in 2001, compared to 600,624 in 1991. Sixteen towns were selected for a detailed comparative case study, using an approach based on the holistic methods of regional geography. Such methods involve analyses of individual natural and human aspects of the small towns (relief, soils, climate, water, biogeographical elements, population, dwelling function, productive branches, technical infrastructure, transport, tertiary sector, tourism, etc.). As part of the research design, factors related to geographical position and historical development were also incorporated. The broad objective of such an approach is to evaluate relations between natural and human subsystems. Some of the typical outcomes from these field-based projects are, for example, a typology of landscapes, natural hazards, relations between a town and its hinterland, microregional territories of the town and their evaluation, environmental problems (ecological stability, protection of landscape, pollution and its consequences, old waste deposits, living conditions, social milieu and pathological features, etc.). Since all of these relations between humans and nature can be interpreted from certain viewpoints, we can infer only a partial synthesis of the collected information. In another part of the first phase of the project, we held discussions with local professionals on their perspectives and future visions for the small towns.

The methodological framework is still under development as the research project continues. For example, in the last three cases examined, survey research methods were added to the fieldwork design. This should lead to a better understanding of the images of small towns and their characteristics, as perceived by small-town residents and by visitors.

Some Principal Findings

Small towns in Moravia have a number of common characteristics, but each presents an individual case of the interaction of historical traditions and efforts by the current generation to plan for future prosperity, all this being set in a particular landscape and location, with particular relations with surrounding villages and towns. In the following discussion, however, some general findings based on the sixteen case studies are presented; for detailed findings for each case, see Akademie věd České republiky, Ústav geoniky, pobočka Brno (2002).[19]

Economic Characteristics of Small Towns

Industry was and still is, in most cases, the most important source of prosperity in the small-town sector. Moreover, employment in industry and power production sectors in Moravian small towns comprises over one third (36.1 percent) of the economically active population, and this represents the most industrial part of the national settlement system. Basically, industry supplies the opportunities for work, supporting the income levels of small-town populations and the local demand for services. As a result, the necessary conditions are created for better education and professional changes in the structure of the small town.

In the socialist period, attention was focused in particular on the development of heavy industry. In a number of the towns, small factories were changed into complexes with thousands of employees, often then out of balance with the town size (e.g., Adamov and Břidličná). Small towns offered employment opportunities to people from villages in a large catchment area, augmented to some degree by workers "freed" by higher productivity in agriculture. Undoubtedly, the third quarter of the twentieth century brought about significant modifications in the structure of small Moravian towns.

After 1989, economic restructuring of the small town sector started. Besides the general change from an input-output economy, a number of factors can be noted: the rapid decrease of employment in industry compared to increases in service jobs; the effects of privatization—often complex, with quasilegal changes of ownership—as well as the fact that privatization was both more rapid and more complete in coverage than in other East Central European countries; problems associated with long-established, "oversized" industries with many branches and outdated technology; and the effects of losing markets after the separation of Slovakia. An immediate result was seen in the numbers of people unemployed—in one third of the case study towns, employment opportunities in industry decreased by 50 percent, sometimes more.

The current situation of the industrial sector in the small towns can be characterized in two main ways. In a number of these towns, the local economy was dependent on one large industrial complex, while other, smaller establishments served as feeders for the giant (e.g., Adamov, Břidličná, and Fulnek). With few exceptions, these establishments have survived, but the number of employees has been reduced. Their future depends on some variant of reindustrialization. For example, the demise of a large light industrial complex (such as that for wood processing in Bučovice) can result in the creation of a number of small establishments, largely with the same expertise and production lines, making use of the qualifications of workers, level of infrastructure, extant marketing networks, and so on. On the other hand, for the metallurgical or armaments industry, the situation is clearly different.

A second general characteristic is that small towns with diversified light industry and agricultural products are in a better situation. Some small towns saw a rapid dissolution of industrial production (e.g., Bzenec), but often a structure based on different branches, size, and ownership is more adaptable to changing conditions. In place of the liquidated factories, there are today new establishments with similar or completely different types of production in these towns.

About one half of the small towns were able to attract direct foreign investments, which is very important, given the lack of national capital. Although the foreign investments offer only hundreds of jobs, as a rule they advance local purchasing ability and bring advanced technologies and organization of labor to small towns.

Foreign investments come to small towns essentially in two ways. The first one is the purchase of a Czech firm within the privatization regime and its restructuring. As examples, one can mention the plants of Dürkopp Adler Boskovice (formerly Minerva), a producer of machines for the leather industry (1,100 employees), or of Siemens (formerly the Moravian Electrotechnical Works), which makes electric motors at Frenštát pod Radhoštěm (1,100 employees). A foreign firm sometimes uses an old area of production—even agricultural space—as a location for its plant, as in the case of MWG Bzenec, a manufacturer of metallic furnishings (200 employees).

The second way is construction of a new plant, usually based on promotion or advocacy by the community. The offer can include parcels of land with technical equipment, legal and organizational assistance, and some other incentives. Bystřice nad Pernštejnem is one example of a town with two foreign enterprises located in its industrial zone: Inter Transtech, a producer of gears (220 employees), and Rathgeber, a maker of name-plates and similar objects (fifty jobs). The Siemens company was so satisfied with its first investment that it built another division: Siemens Automobile Engineering, at Frenštát pod Radhoštěm (another 1,200 jobs). The existence of a qualified but relatively cheap labor force is often the motivation for foreign investment. One such startup is the Japanese plant ALPS Electric at Boskovice with 400 employees.

Current Employment Status

At best, the general situation in terms of employment can be classified as "average" in the Czech Republic, but in most cases it is worse than that. The tertiary sector (services) has shown significant growth, even in small towns, but not to the extent of absorbing all of the available labor force freed from industrial employment. The lowest unemployment levels exist in towns in close proximity to Brno, as well as in some small towns with their own significant economic activity. A very high degree of unemployment is found in small towns in marginal locations and in those towns where a large industrial complex has disappeared.

Measures of unemployment, however, are not only a ratio of "jobs sought" compared to "jobs available." Some poorly established criteria in social policy have created relatively large groups of unemployed but highly qualified people, who would rather accept social support or welfare (with perhaps some extra money received from occasional illegal activities) than earn low salaries for less qualified work. This causes the paradoxical situation of high levels of unemployment coexisting with a scarcity of labor that leads to hiring workers from Slovakia, Ukraine, Belarus, and possibly some other places (as, in Bzenec). On the other hand, some towns have ample job opportunities for qualified residents and, at the same time, a lack of labor for existing work openings in light industry (e.g., Brtnice).

In the socialist era, the tertiary sector in small towns lagged behind the West European situation. This applied less to the social sector (education, health, social services, etc.) than to commercial services. Compared to village settlements, small towns held exceptional positions as a result of the "central settlement system" of administrative planning. After 1990, however, there were significant movements of workers from the secondary to tertiary sectors, in contrast to the very slow rates of movement in the socialist era. These changes also brought about certain problems, including people's inability to undertake entrepreneurial activities to any serious extent, the absence of a financial infrastructure, and an ineffective legislative process. Even local demand in terms of disposable income was missing. In a context of attitudes held over from collectivization, there was also a low level of cooperation between individual enterprises. The small towns had several disadvantages in competition with the larger cities, partly as a result of the expansion of individuals' transportation options. As rates of car ownership have increased, many consumers prefer to travel longer distances to shop in large hypermarkets (such as Tesco, Hypernova) in Brno rather than in the traditional stores in the small towns.

The tertiary sector may also serve to improve the future prosperity of small towns, especially in cases of fully developed centers with a hinterland. About one third of the set under study belongs to this category of towns: Boskovice, Bučovice, Bystřice nad Pernštejnem, Bystřice pod Hostýnem, and Frenštát pod Radhoštěm. The most important factors for the success of these tertiary functions include the size of the population in the local system, the centrality of the town, various social factors (e.g., the cultural background of the residents), some specific conditions (e.g., cultural/natural amenities), and, especially, human factors. Services, more than any other sector, depend on customer satisfaction, almost exclusively a function of the human factor.

The human factor should be included also as one of the indicators of the quality of life, and in this sense, the level at which the local population can be entrepreneurial is important. In part, such attitudes can be attributed to cultural and entrepreneurial traditions and the stability of social conditions, and to the educational structure of the population, in terms of qualifications. A less rich cultural environment, for example, may have been created in those towns where there was a significant movement of people after World War II either because of industrialization (e.g., Adamov and Břidličná) or ethnic exchange of people (e.g., Budišov nad Budišovkou and Fulnek). One could also include those small towns that until recently had high levels of agricultural employment. The quality of life and social environment in individual towns can also reflect activities of local community groups and organizations.

Because some types of services such as those in the social sector (educational services, health, social services, etc.) need not be located at higher-level centers, some degree of specialization in the small towns could be developed. The quiet environment of small towns could be an ideal location for these services. At the same time, such services would bring qualified workers into the town, raising the intellectual and cultural level of the population. Furthermore, if these centers were to be governmental, or subsidized by government, they would also bring financial support to the town. Specialization in tourism is another possibility.

Tourism

Many expert studies have suggested that increased prosperity for marginal regions (including small towns) could be brought about by the development of tourism. At present, this approach does not appear to be realistic for Moravian towns. Though the majority of them have historical, cultural, or natural attractions, the heavy industry that was established in some of these small towns negates the quaint and unspoiled atmosphere necessary for recreational centers as one knows them in the Alps, or the seaside resorts in Western and Southern Europe. The inhibiting factor is not the simple presence of old industrial sites, many of which are rarely active any more. Rather, one has to take into account the (non) availability of funds for reconstructing the buildings, the lack of professional persons to undertake the work, and the prevailing attitudes of the people, who used to work in the industries.

Another drawback afflicting the sector is the low availability of tourist services. The small towns are geared to serving a relatively undemanding clientele. In the tertiary sector the small towns can compete with each other, but for tourism a better option is to cooperate—to formulate "packages." At the moment, there are few established ways to do this. Some exceptional cases do exist, however, marketing the "tourist region" to both local residents and outside visitors. One is the so-called Vlachian "kingdom," an association of entrepreneurs in Eastern Moravia. As for the potential international clientele, the current problem might be the fact that they know of Prague and a few other places in the Czech lands, but small towns in Moravia are as yet of little interest to them.

Quality of Life/Natural Environmental Factors

An important competitive advantage of the small towns is their living conditions. Over the last decade, there has been noticeable progress in terms of technical infrastructure. The resulting improvements in living standards have largely counterbalanced the former negative factors. The most obvious changes are the switch from coal and solid fuel to gas, the improvements in sewage systems, and legislated and organizational changes in solid waste management. Compared to large towns, small towns have the advantages of a higher-quality social milieu together with closer contacts with the surrounding natural landscape. With the exception of those towns where new apartment blocks were constructed to house the workers for the industrial complexes, small towns enjoy better living conditions, characterized by higher percentages of single-family houses and more actual living space.

As anywhere, the most important environmental problem in the small towns is increased traffic, especially at major highway intersections (about one third of the case towns; e.g., Bučovice, Bzenec, and Fulnek). In some cases, the quality of the social environment is at risk because of higher unemployment, which can be a catalyst for other social pathologies; one can even witness problems like drugs and criminality in the small towns, albeit on a smaller scale than in the big cities.

More generally, the attraction of small-town living conditions could derive, in part, from their demographic situation. Small-town population increases result from combinations of both natural factors (fertility, mortality) and the effects of in- and out-migration. For the industrial centers in the set of small towns, which have experienced high levels of in-migration, the gains are due to the younger base population with its higher fertility rates. At the same time, living conditions in such centers are not very attractive, so out-migration is increasing. On the other hand, an interesting comparison can be made to those towns that were bypassed by industrial development. They have an older population structure that is experiencing a natural decrease in numbers, but at the same time they can today be destinations for in-migrants because of their generally more attractive settings and environment.

The Future of Small Towns and Sustainability

Given the previously noted demographic and social trends, it is clear that small towns offer certain elements of the Moravian population an alternative to life either in large cities or rural villages. We have established already several differences between the case studies, aware that such differences could be place-specific. Yet, as a set, there are certain generalizations that appear to be plausible and useful in thinking about possible futures. How, then, can we characterize these alternatives, indirectly addressing questions of sustainability in this manner?

Sociocultural Alternatives

Small towns are differentiated from other forms of settlement by a typical way of life that combines an established basic urban infrastructure with a relative absence of problems caused by higher concentrations of activities. One illustrative result is the options for movement around the town—there is no need for a car. The citizen who travels by foot or bicycle experiences the town center and its citizens very differently from one driving a car. The small-town resident has face-to-face personal contact with other residents and an immediate perception of all activities in the locality. One of the factors contributing to this higher level of social life and control is the personal security experienced in small towns. There is a constant, immediate control exercised by citizens in that any outsider is immediately a focus of attention. One physical proof of this is the surviving tradition where people in small towns greet each other—*"Dobrý den."*

The cultural life of small towns can be very different from that of other settlements. While the large cities are open to the globalization of professional culture with a large selection of various activities, in the small towns there is a great deal of local culture. These cultural activities are organized by local entities and, to a large extent, influenced by local citizens on an amateur basis. This does not imply that small towns have lower levels of culture. Almost everywhere there are handicraft schools, amateur groups (e.g., for theater and music), and organizations for various interests. Without undue emphasis, one might even conclude that small towns are important contributors to sustaining national and regional culture, as they are putatively more resistant to the increase of commercial and American culture.

Whenever the context of globalization and its effects on national culture have been discussed, a number of experts have reached the conclusion that globalization is a "natural" process deriving from technological development, and that it could have a positive effect on national culture in the sense of rapid expansion of new ideas and technologies. On the other hand, it would be harmful if globalization destroyed local, regional, and national identities. It is evident that small towns have a solid predisposition for sustaining local identity and, therefore, the identity of the region. One manifestation of the small-town advantage in this regard in contrast to bigger cities is the practicality of holding reunions of people who were born in the locality but have since moved away. Through such reunions, the intellectual and economic potentials of these native sons and daughters may be fed back into local community preservation and development.

Sociopolitical Alternatives

The form of governance and decision making in the small towns is also different. In large cities, candidates are elected from different political parties, but in small towns the basis for choosing officials is more likely personal knowledge, relatively divorced from party politics, and candidates can often be considered independent. One primary implication of this phenomenon can be seen in the way in which people resolve certain conflicts in administration, compared to a city with its ideologically colored fights and competing interests. Better contacts between people in administrative and professional positions and town residents may result from this more informal process, even though one could argue that similar conditions are equally apt to lead to rule by a "small-town elite."

Locational Alternatives

The functional differentiation of Moravian towns is clear, but with respect to the futures of small towns from a locational perspective, there are several unknowns. Some sets of possible futures appear likely, however:

1. Residents of small towns in the hinterland of a large city, i.e., Brno and Ostrava, can satisfy their expectations for employment and services in the large cities. Such towns are little concerned about the danger of economic decline; the largest risk is loss of identity in the face of the virtual impossibility of retaining their individuality within the agglomeration. Such a danger is lessened in situations when the small town satisfies some basic function, usually in cooperation with the larger town (e.g., recreational possibilities).
2. Polyfunctional small towns, which are centers for rural hinterlands, most frequently located in marginal areas with no larger cities in proximity. Today, the overlapping of their micro-regional hinterlands will not be well defined, as they are far enough apart to have low levels of between-center competition. Their role will be mainly to supply basic services and to offer venues for social contact to relatively large but not very wealthy hinterlands. Most of the time their identity will be the same as that of the microregion.
3. Small towns with no obvious hinterland but with functional specialization in terms of the distribution of work. Until recently, the only important specialization has been industrial activity. Yet there could be some future specialization in health care, education, and other social services (if these towns have certain locational advantages, as discussed earlier), as well as in tourism.
4. Ruralized small towns cannot attract any of the successful transformational factors, and in the future they will tend to be downgraded to the status of villages. Today, representatives of these towns are promoting living conditions as an attraction to potential residents, but if they are not on a rail transit link or some other easy transport connection their future will remain highly questionable. One possibility is that such towns could prove attractive to older persons as retirement communities.

Concluding Remarks: The Future of Small Towns in Moravia

The importance of small towns for the development of Moravia can be seen on many different levels. The relatively equal distribution of population noted at the outset should be maintained, as it is an important factor in the retention of villagers in marginal regions. Rural residents will stay there only if supported by basic services, employment opportunities, and social contacts in the small towns. Thus, the relative equalization of the population in locational terms is important, as much for maintaining the country as for indirectly encouraging people to stay in the rural areas rather than following the "natural path" to the city.

Most small towns play a very significant role by offering an alternative life style and enabling the continuing development of society as a whole. The better conditions of the living spaces, the feeling of security, the social connectedness, and the slower pace of living are critical desiderata in a modern world characterized by a hectic pace and impersonal relations. Meanwhile, closer contact with nature and the higher ecological stability of the natural surroundings of small towns are important environmental factors of sustainability.

Even if small towns do not and will not play the most important role in the future development of the economy, their stake in the continuing development of the economy is not negligible. The current share of small towns in industrial production is still important. It can be hypothesized that, while the large cities are set to undergo development based on the communications technologies connected with the quaternary sector, light industry could move into small towns, where it could also work with local agriculture and agriculture-related industries.

Indeed, it can be asserted that the continued development of the Czech Republic will not be possible without the small towns. Viewed optimistically, it is likely that there will be some future sharing of objectives between larger cities and small towns, in which the large cities are sources of development stimuli contributing to the evolution of the country, while the small towns continue to provide a cultural base reflecting more permanent values.

To incorporate fully the "new realities" of both economic forces and sustainability at a global level and to find appropriate transformations to regional pathways for development, any investigation must account for the critical issues of governance, civil society, and the regional context in which they have signification.[20] In the case of Moravia, we have argued that the small-town sector can serve a vital role in this respect. For East Central Europe in general, some of these findings may have some degree of applicability, owing to similarities in the experiences inherited from the socialist era. It is clear, however, that future research must attend to both the theoretical frames for investigating the so-called transition or transformation[21] and the empirical situations appropriate for any study of regional political economies. Otherwise, the separation of environment and economy will continue to confound even the short-term promises of sustainable development, particularly in the face of accession to the European Union. Even with greater attention paid to environmental effects under the sustainable development scenario, one can expect only marginal, incremental improvement of the situation.

But sustainability, writ large, is different. Culture, politics, people ... the context for relations between "economy" and "environment" is significantly changed, and for many observers, the sociopolitical milieu is paramount. Since any movement toward a sustainable society requires rewriting the ground rules,[22] the changed orientation is inevitably political because it is proactive. Certainly, individual localities, such as the small towns presented in this chapter, can work progressively in this way, and perhaps the recent revival of bioregionalism augurs well for regional groups of communities.[23] After all, people make "region" as much as they make "sustainability."

Notes

1. This paper was produced as part of the "Geography of Small Towns" project funded by the Grant Agency of the Czech Academy of Sciences Nr. A3086301.
2. A. J. Scott, "Economic Geography: The Great Half-Century," in *The Oxford Handbook of Economic Geography*, ed. G. L. Clark, M. P. Feldman, and M. S. Gertler (Oxford: Oxford University Press, 2000), 18–44.
3. P. Cabus, "The Meaning of Local in a Global Economy: The 'Region's Advocacy of Local Interests' as a Necessary Component of Current Global/Local Theories," *European Planning Studies* 9 (2001): 1011–29.
4. B. Greer-Wootten, "Towards a Geography of Sustainable Development and/or Sustainability?" *Moravian Geographical Reports* 12 (2004): 2–9.
5. N. Brenner, "Globalisation as Reterritorialisation: The Re-scaling of Urban Governance in the European Union," *Urban Studies* 36 (1999): 431–51.
6. B. Greer-Wootten, "Sustainability and Scale," in *Regional Prosperity and Sustainability*, ed. P. Hlavinková and J. Munzar (Brno: Regiograph for Geokonfin, 1999), 75–85.
7. A. Fagin, "The Development of Civil Society in the Czech Republic: The Environmental Sector as a Measure of Associational Activity," *Journal of European Area Studies* 7 (1999): 91–108; A. Tickle, "Regulating Environmental Space in Socialist and Post-socialist Systems: Nature and Landscape Conservation in the Czech Republic," *Journal of European Area Studies* 8 (2000): 57–78.
8. Scott, "Economic Geography."
9. G. L. Clark, M. P. Feldman, and M. S. Gertler, eds., *The Oxford Handbook of Economic Geography* (Oxford: Oxford University Press, 2000), viii.
10. Ibid.
11. P. Hall, S. Marshall, and M. Lowe, "The Changing Urban Hierarchy in England and Wales, 1913–1988," *Regional Studies* 35 (2001): 775–807.
12. P. Claval, "Réflexions sur la centralité," *Cahiers de Géographie du Québec* 44 (2000): 285–301.
13. J. B. Parr and L. Budd, "Financial Services and the Urban System: An Exploration," *Urban Studies* 37 (2000): 593–610.
14. T. N. Clark, "Old and New Paradigms for Urban Research: Globalization and the Fiscal Austerity and Urban Innovation Project," *Urban Affairs Review* 36 (2000): 3–45.
15. A. Dingsdale, "Central Places and Planning in Hungary," *Geography Review* 6 (1993): 12–16.
16. J. Hinderink and M. Titus, "Small Towns and Regional Development: Major Findings and Policy Implications from Comparative Research," *Urban Studies* 39 (2002): 379–91, here p. 389.
17. See, for example: A. Borsdorf and M. Paal, *Die Alpine Stadt zwischen lokaler Verankerung und globaler Vernetzung* (Vienna: Verlag der Österreichischen Akademie der Wissenschaften, 2000); E. M. Munduch and A. Spiegler, *Kleinstädte: Motoren in ländlichen Raum*, Landtech-

nischen Schriftenreihe, no. 214 (Vienna: Österreichisches Kuratorium für Landtechnik und Landentwicklung, 1998); M. Niedermayer, *Kleinstadtentwicklung* (Würzburg: Geographisches Institut der Universität, 2000); M. Perlik and W. Batzing, *L'Avenir des villes des Alpes en Europe* (Bern: Verlag des Geographischen Institutes der Universität, 1999); and F. Žigrai, "Niekoľko poznámok k problematike malých miest," in *Urbánny vývoj na rozhraní milénií*, ed. R. Matlovič (Prešov: Filozofická fakulta univerzity, 2000), 180–2.
18. A. Vaishar and E. Kallabová, "Vývoj služeb v malých moravských městech po roce 1990," *Geografie* 106 (2001): 251–69; E. Kallabová, "Changes of Hierarchical Positions of Small Moravian Towns on the Example of Education," in *Nature and Society in Regional Context*, ed. P. Hlavinková and J. Munzar (Brno: Regiograph for Geokonfin, 2001), 74–82; A. Vaishar and J. Zapletalová, "Jemnice: The Role of a Small Moravian Town in the Present Stage of Transformation," *Moravian Geographical Reports* 6 (1998): 32–42; A. Vaishar, P. Hlavinková, S. Hofírková, M. Hrádek, E. Kallabová, K. Kirchner, A. Klímová, J. Lacina, S. Ondráček, E. Quitt, J. Škrabalová, B. Trávníček, and J. Zapletalová, "Geography of Small Moravian Towns: Case Study Bučovice," *Moravian Geographical Reports* 9 (2001): 43–62.
19. Akademie věd České republiky, Ústav geoniky, pobočka Brno, *Geografie malých moravských měst I* (CD-ROM) (Brno: Ústav geoniky, 2002).
20. D. Molnar, A. Morgan, and D. V. J. Bell, *Defining Sustainability, Sustainable Development and Sustainable Communities* (Toronto: Sustainable Toronto Project, 2001).
21. E. Altvater, "Theoretical Deliberations on Time and Space in Post-socialist Transformation," *Regional Studies* 32 (1998): 591–605.
22. L. R. Brown, *Building a Sustainable Society* (New York: W. W. Norton, 1981).
23. M. V. McGinnis, *Bioregionalism* (London: Routledge, 1998).

Chapter 15

BUILDING LOCAL SUSTAINABILITY IN HUNGARY

Cross-Generational Education and Community Participation in the Dörögd Basin

Judit Vásárhelyi

The State of the Environment and of Local Communities in Hungary

From the point of view of sustainability, Hungary's pattern of settlement has an unhealthy structure. Two million of the country's approximately 10 million inhabitants live in the capital, Budapest. Of the remaining four fifths, some 4 million people live in other cities and towns, but the second largest to Budapest is smaller by an order of magnitude. Approximately 3.8 million people live in small, rural communities.

The state of the environment in the various settlement areas is partly the heritage of the command economy, which did not reckon with environmental quality in its production goals. Despite the substantial reductions achieved in emission levels of pollutants such as sulfur dioxides, nitrogen oxides, dust, and greenhouse gases, air quality is still unsatisfactory in most urban areas (the agglomeration around the capital and the industrial towns). After the fall of communism, significant reductions in industrial pollution were offset by steadily increasing pollution from individual and commercial transport. In fact, by the end of the 1990s, transport had become the major source of urban air pollution in Hungary.

Because of its location at the center of the middle Danube basin, Hungary has the highest annual average per capita surface water flow in the world. However, pollutants flow in across the borders, and local municipal and industrial entities lack adequate sewerage and other wastewater treatment facilities. Together these factors have made about three fourths of Hungarian surface water resources vulnerable to pollution. There is also significant pollution of groundwater, mainly fertilizer nitrates and traces of pesticides from industrial farming.

Hungary's most valuable natural resource is *soil:* some 85 percent of its territory is suitable for agricultural activities. The greatest problems here are the lack of a comprehensive strategy for protection of the land and the deterioration in the quality of the land. Many rural settlements suffer from decay and deterioration in the quality of life, due to insufficient economic activity, migration, and changes in the demographic structure of the countryside. Furthermore, the abandoned industrial areas in rural communities are not being rehabilitated, there is widespread littering and neglect of public areas, and green areas have not been extended to the limits prescribed by law. There is growing fear that rural settlements with traditional features have little chance of survival. Every year, a number of small and medium-size settlements representing unique values, traditions, and imagery are either eliminated or find themselves fossilized as country museums. Thus, despite some improvements in environmental quality in major urban areas and some indications of progress toward sustainability, many rural communities are still under the threat of virtual extinction and far from developing a sustainable economy.

Opportunities for and Threats to Sustainability in Local Communities

After the political and economic transformation in 1990, the freely elected Hungarian parliament enacted a code that gave local governments decentralized rights of decision making. However, neither this autonomy nor the numerous new tasks associated with it have come hand in hand with the necessary financial resources. Soon, the euphoria over self-government vanished as it was realized that both the local governments and the people themselves sorely lacked the skills and experience needed to solve old and new problems, to develop workable scenarios for local sustainability, and to raise the funds necessary for such programs. They could not even recognize that unsustainability and limited local resources were their particular problems in the new economic context following the collapse of socialist industry, in which multinational firms were arriving and the hitherto unknown phenomenon of unemployment had appeared.

Therefore, there was an urgent need to fill that gap in knowledge and experience by involving a variety of social groups and stakeholders into a historical learning process: designing a collaborative planning process for their own communities, building consensus, maintaining democratic decision making, and securing a high level of regular public participation. Many initiatives throughout the country, including local environmental action plans (LEAPs) and community action programs (CAPs) focused on promoting local sustainability. The initial and most fundamental step was to develop ways to educate people in the various aspects of sustainability. The main idea was to design curricula that would bring together members of different generations within the community, as well as representatives from a variety of economic sectors, to participate in the same training exercises at the same time. In this way, the idea of sustainability could be introduced into daily practice and start to influence the behavior of both producers and consumers.

One of the first activities undertaken in preparing local action plans was to assess local resource potential and prospects for self-sustaining activity. The local planning community had to identify sustainable uses of local human, land, and other resources and start planning future developments within these parameters. They had to create a capacity for managing the local resources as well as to strengthen local identity in the midst of changes and transformations. Finally, they had to cope with the lack of political will and coordinated programs at the national level regarding the implementation and support of programs in local sustainability.

Local Sustainability: National Legislation and European Assistance

Shortly after the Hungarian delegation came back from Rio de Janeiro in June 1992, the government passed a measure (Nr. 1024/1993) on the tasks related to the conventions adopted at the Earth Summit. It also adopted Agenda 21, which made all ministers responsible for submitting concepts of "concrete possibilities" of implementing the international agreements. Though smart reports were sent periodically to the UN, so far, no clear concept of sustainable development has been elaborated or implemented for Hungary as a whole. Meanwhile, some of the environmentally positive results in the reports—such as the trends toward decreasing use of either energy or chemicals—were actually symptoms of the difficult situation of the economy in the transitory period.

Just recently, however, direct measures aiming at sustainability have appeared in governmental programs. In 2001, the government announced programs for energy efficiency, schemes for the utilization of thermal waters, and the national agrarian-environmental program (NAKP). The NAKP was based on a professional interpretation of potentials for local sustainability combined with the principles embodied in the Common Agricultural and Rural Policy for Europe (CARPE) or the proposals for reform of the Common Agricultural Policy (CAP) of the European Union. In 2002, Hungary's Ministry for Agriculture and Regional Development even launched an operational program for NAKP. Within the framework of this horizontal, country-wide program small farmers, based on a five-year contract, voluntarily agree to undertake environmentally sensitive cultivation of their land, thus limiting the intensive industrial farming methods that undermine sustainability. Some of the contracts include obligatory training on the methods appropriate to sustainable agricultural land cultivation.

Some EU pre-accession programs, like SAPARD (Special Accession Program for Agriculture and Rural Development), could serve as suitable vehicles for channeling local Agenda 21 approaches into the local development strategies. So far, more than two hundred strategies for small communities have already been evaluated. Some of them, initiated by environmental NGOs, are focusing on the protection of the highly valuable ecological resources of their respective territories. It is worth mentioning that these strategies are based on careful planning and have secured real public participation. Despite these positives, the actual start of the operation has been postponed awaiting a decision on concrete support from SAPARD.

Learning from the Past: The Commons—A Tradition of Sustainability

There was a long tradition of measures at the village level regarding the sustainable use of commons—forests and pastures—which was coded into sophisticated written as well as oral rules that go back to the seventeenth century. Unfortunately, this great tradition has almost disappeared in the Carpathian basin. Only very old men remember the practice. The Hungarian practice was researched by a historian, who uncovered old records from Transylvanian villages and found that the rules related to the use of the commons (and serious sanctions for violations) were enforced equally for everybody, without discrimination according to social rank in the village hierarchy. Both the landowner and the peasants accepted the decisions of the community on a year-by-year forest management plan, which involved strict rules providing for the maintenance of the forest economically and ecologically. The forest was protected both quantitatively and qualitatively, while room was made also for satisfying the community's need for wood.

This old tradition cannot be restored directly in contemporary society, but its message should be adjusted to new conditions. It could then contribute to a further articulation of the concept of sustainability and gain influence in shaping the mentality of both rural communities and organizations such as the Society of Foresters and the Hunters Association, which are active in rural areas. Politically, the notion of sustainability could be shown to resonate with the national historical tradition.

Examples of Local and Regional Initiatives for Sustainability

There are a number of programs and initiatives in Hungary that do, by and large, conform to the essential ideas of sustainable development. The program for the Tisza Valley, the strategy and physical planning for the Körös-Maros region, and the development concept for the Ráckeve-Soroksár region, all initiated by the government, share characteristics such as being in harmony with the recommendations of the United Nations Conference for Environment and Development (UNCED). They also show a strong process of participation involving various stakeholders, including NGOs.

However, the issue of local sustainability has been raised more frequently by national and local NGOs. One of the most interesting models is the comprehensive Sokoró Region Initiative, which promotes a deliberate reorientation toward traditional yet sustainable agricultural methods that adjust cultivation and production to natural conditions; for example, growing fruit trees rather than wheat on the hillsides. The reintroduction of native fruit species also has contributed to the development of eco-tourism, with programs like demonstration of renewable energy, fruit-tree gene banks, the breeding of traditional horses, and the maintenance of natural wild animal areas and education trails reducing local unemployment, all in all. The non-profit Sokoró Foundation together with the local government have established a firm, which provides expertise and offers a franchise-type production model to the farmers.

Other examples of successful local actions can be given. Within the traditional small villages of Ormánság, surging free-market capitalism threatened what was left of the natural resources after the command economy. The Ormánság Foundation started its program by emphasizing the autonomy of the landscape. Protection of the landscape heritage was combined with a solid technology program and community protection. In the small village of Gömörszöllős with 199 souls, the local NGO, the Ecological Institute, a foundation for sustainable development, did a comprehensive survey and took responsibility for the social situation, economic activities, and the state of the environment there. Their program includes many simple alternative technologies, training and consultation, and the practice of sustainable tourism and agriculture. The 360-hectare Galgafarm First Hungarian Organic Agricultural Cooperative was also initiated by an NGO. Despite initial difficulties, it was selected as a member of the European Network for Sustainable Development of the European Parliament. Finally, the Somogy Nature Preservation Organization was the first NGO in Hungary that purchased land with a view to nature preservation. This organization also was entrusted with managing a protected area—the first time that a civic organization in Hungary has been assigned a state task.

Environment and Development in the Dörögd Basin: A Case Study, 1991–2002

The Dörögd Basin is a rural catchment area bordered by five small villages, partly overlapping the Lake Balaton Uplands National Park, in the Transdanubian (western) part of Hungary. The Independent Ecological Center (IEC), a Budapest environmental civic organization with several partners and sponsors, started a rural community program here as early as 1991. The interactive process was based also on the experience of local environmental action plans (LEAPs) in fifteen small towns in Hungary (1992–1997). The method of comparative environmental decision making that was used here was introduced in Central and Eastern Europe by the Institute for Sustainable Communities (ISC) of the U.S. state of Vermont, a partner of the IEC. The procedure was adapted to Hungarian conditions. The IEC aimed to use the lessons learned in the urban communities to broaden its rural project, targeting depressed rural settlements that nevertheless possessed valuable natural resources and local traditions. Since 1991, the IEC has been running an ever widening program for organic development in the Dörögd Basin. In the course of the activities, from the first community college (Volkhochschule) through the organic development phase, knowledge on local issues has been collected. The small region has a number of natural and cultural values.

The Landscape

The earliest archaeological finds in the Dörögd Basin that demonstrate human activity date to the ninth century B.C. The soils preserve the memory of forest masses. The landscape was significantly changed by the activities of the Roman Empire, which built military highways and settlements along the Eger-víz (the Eger stream). The medieval settlements were smaller: fourteen villages shared the area, separated from each other by forests. During Turkish rule, the majority of villages became depopulated, and at the turn of the seventeenth to eighteenth century only six villages were repopulated. These villages continuously increased the acreage of their fields, so that first the forests, then the pastures disappeared from the center of the basin and from the peripheries of the villages. At the end of the nineteenth century, the fields not only climbed up onto the slopes but also claimed the wetlands. However, hillside farming began to decline with the spread of large-scale intensive farming of monocultures.

The landscape contains several areas or features that have been designated for protection due to their unique natural or cultural values. The **Balaton Uplands National Park** was established in 1997. Park lands stretching into the basin include the Királykő cliff at Kapolcs, the wetland area around the Kálomisz Lake, and the Imár Hill, the central feature of the basin. All of the basin's five villages claim this landmark as their own. The inhabitants have erected a wooden cross on the ridge of the hill. The buzzing of the bees and the scent of flowers mixed with this holiness all go well together with the nearby fire-place, where locals celebrate the midsummer's night. Numerous protected plants can be seen flowering in the spring wind. One of the most extraordinary of these is the abundant, lemon-scented, thickly flowering burning bush. The Király-kő, a steep, gray, 400-meter-long basalt formation, visible from all look-out points, can also boast of botanical rarities. Some other elements of the landscape are protected locally. The Nagy-tó (Big Lake) and its beautiful surroundings are the habitat of several rare species. The most valuable part is the peat moss that covers approximately 35 percent of the area. The Cloister-Spring and the Tálod Monastery ruins can be found in a quiet woodland setting. The Eger-víz crosses the area. A small section of the living landscape is supported by famous natural springs.

Geology

Very few of the minerals found in the Dörögd Basin are worth collecting: olivine, calcite, aragonite, barite, and manganese dendrite. Quarries were opened to produce building materials: stone, sand, clay, gravel and lime, or dolomite; their sites can also be found in the area. The basin also has deposits of a rare mineral called alginite, made up of algal biomass and volcanic tuff disintegrated into clay, which is used to increase the fertility of the soil. The diversity of the basin's fauna is due partly to the numerous and varied kinds of rock here.

Climate

The climate of the basin shows an interesting balance of continental, Atlantic, and Mediterranean effects. The amount of precipitation to fall on the land exceeds the amount that evaporates. Consequently, the natural vegetation of the area is forest. The snow cover lasts from November till mid March. Summer's frequent cloudbursts are a significant erosion factor. The dominant wind direction is north northwest.

Water

The vital element for life, water, is found in many forms in the area. On the surface, water primarily occurs in the form of springs, streams, or lakes. As precipitation it can come as rain, sleet, hail, dew, fog, frost, snow, and rime. Beneath the surface, clean cavern water is found in the cracks and crevices in limestone and dolomite.

An enormous amount of water extraction following bauxite mining in a neighboring area—a former Soviet interest—coincided with intensive deforestation and years of drought, causing water to disappear in the basin for almost a decade. Many springs, streams, and lakes fell victim to man's thoughtless interference. Several lakes have formed on the beautiful basalt plateau. The most significant surface water feature is the Eger-víz, a stream that crosses the basin. The entire network of the basin's streams belongs to the catchment area of Lake Balaton.

Vegetation

The vegetation in the basin is diverse and colorful, since the geological surface is also extremely varied. The decade-long disappearance of the region's streams led to significant changes in the conditions for vegetation, including the wilting and slow demise of waterside plants. The most characteristic forest types are the Turkey oak-oak and the hornbeam-oak forests, although extrazonal beech forests are also to be found. Abandoned, ungrazed pastures and uncut meadows have become overgrown with shrubs and trees: the blackthorn, hawthorn, dogrose, and barberry are the first to become established.

Botanically, one of the most valuable pastures is Imár Hill, which is surrounded by fields at the junction of the five village boundaries and has never been forested during its use. Many historic events and legends revolve around this place. The area was put first under local, then national oversight of nature conservation. Varied types of habitats have evolved to accommodate more than two hundred plant species, amongst them the purple anemone and small pasque flower, mottled iris, yellow adonis, silky buttercup, maiden pink, and burning bush.

Animals

Some prehistoric finds tell us about the animals that once lived here: for instance, the shattered remains of a mastodon were found in the Pannonian (lower Pliocene) layers.

The varied relief conditions and the vegetation of the Dörögd Basin enables mammals requiring different habitats to live in proximity to each other: the stag, the boar, the fox, the badger, the marten, the hare, the squirrel, the large dormouse, the mole. One can meet roe-deer, even during the daytime, both in the forest and in fields. The last colonies of the legally protected gopher, considered an agricultural pest only a few decades ago, can still be seen on the highest point at Öcs.

Many bird species are at home here: the turtle dove, the common swallow, the house martin, the corn bunting, the yellowhammer, and the buzzard. The goshawk hovers, hunting for grasshoppers, insects, voles, and mice in the grass. The red-backed shrike, the Old World skylark, and the crested skylark are quite commonplace. One can also meet the great spotted woodpecker, the serin, the magpie, the golden oriole, the white wagtail, and the lesser whitethroat, and hear the nightingale. Two rare bird species are known to nest here: a pair of bee-eaters and a pair of common wheatears. The black redstart and the linnet are also to be seen only here. The meadows are full of the familiar "pitye-palatye" sound of the quail. Walking quietly in the forest, one can meet robin, wren, chiffchaff, jay, blackbird, chaffinch, the common sandpiper, the song thrush, the woodpigeon, and the mistle thrush. Old orchards are visited by wrynecks and hoopoes.

The region is also rich in reptile, amphibian, fish, arthropod, and insect species.

Energy Resources

As late as the 1950s, energy consumption in the characteristically rural environment of the basin was at a level that could be sustained over a long period. Around the house, people used wood for heating, hot water, and cooking and they used paraffin candles for lighting. In agriculture, they used animal traction for land cultivation. In industry, water mills were used for power and animals for traction.

Today, according to a survey, energy use has increased and diversified. For household needs, the local people use coal, local wood, oil, electricity, and PB gas for heating; PB gas and local wood or electricity for cooking; electricity, local wood, and PB gas for heating water; and electricity for lighting. In agriculture, there are tractors with internal combustion engines for soil cultivation, but animal power also is used. Our research identified a significant potential for use of alternative, renewable energy sources, including solar, wind, and biomass.

The Villages

The structure and position of the Dörögd Basin's five villages is closely related to the assets and the history of the landscape. Öcs was built at a crossroads. Pula was for a long time the most closed-off village in the basin. The mills of Kapolcs along the abundantly flowing stream and the paucity of its agricultural land had made it early on an industrial and trading village. Vigántpetend once fit the pattern of the "one-street village." Taliándörögd was built along the route to the market at the small town nearby. The villages are defined by their architectural traditions as well, especially their stone buildings with rendered (plastered) walls and yards encircled by plain stone walls. The traditional lifestyle was reflected in the settlement pattern of family houses situated within a garden plot. Every house had an ornate flower garden, an orchard, and a vegetable garden supplying homegrown foodstuffs.

Wandering around the landscape, one notices traces of old settlements, the ruins of buildings, bridges, weirs, and mills that give clues to the shape of human presence here in former times. The remains include the thirteenth-century Gothic church of Saint Andrew or another monastery belonging to the old Paulite order. The milling industry used to be a prominent feature of the landscape. This is reflected in many literary artifacts concerning mills and the millers, including documents dating as far back as the beginning of the thirteenth century, sayings, legends, and modern prose and poetry, which also celebrate the other beautiful aspects of the landscape.

The Local Community

Despite the decades of measures taken by the communist regime to discourage traditional small settlement communities, each village within the basin has its own identity, rich spiritual resources, and traditions. The still living crafts in the basin include carpet and broom making, basket weaving, woodcarving, stonewall building, beekeeping, wheel making (cartwrights), the making of corn-dollies, barrel making (coopers), blacksmithing, and shepherding. The locals know how to make cheese, cottage cheese, and toys, how to carve the handle of a knife, bake bread, bind sheaves and bushels, pluck goose feathers, collect and dry wild herbs, and make honeybread—activities that they pursue in addition to their more mainstream sources of livelihood. In Kapolcs, blacksmithing can be seen at first hand at an old forge resurrected as a living history museum, where occasional craft demonstrations are held.

The religious denominations in the area include Roman Catholics and the Hungarian Reformed Church (Calvinists). Only one clergyman lives in the Dörögd Basin; others, who work in several villages, come in from the outside to conduct their services. Each village in the basin celebrates its own patron saint on a feast day called the fęte. This day is a holiday for nonreligious and non-Catholic families as well. It has become a custom for those who have moved away to pay a visit home at this time; cooking and baking go on for days in anticipation of their reception.

Other festive traditions also are still alive here. In Pula there is fasting from Palm Sunday through Easter week; on Holy Saturday people make noise with clappers instead of bells and receive eggs in exchange. Pigs are then slaughtered after the "Swabian" fashion. Of the customs that have survived intact in the basin to this day, the most important is the "green branch walk" on Easter Sunday. On the day of Corpus Christi, a gateway of flowers is built along the route of the procession. Like elsewhere, at Christmas it is a custom for carol singers to enact the nativity events throughout the village. In Taliándörögd they erect a Maypole. At the end of September there is a grape harvest procession.

Typically, the age structure of the population in the villages is older than the national average. There is only one school, housed in five buildings divided between two villages, which serves the children of these plus a third village. The children of the remaining two villages attend schools outside of the basin. The life of young people here differs in many ways from the lives of others living in similar small villages outside the basin. During the summer season, they all assist at the local cultural events. The local youth are recognized through young spokespeople, one from each village, who represent their generation at the fora held in the basin.

The local civic organizations were reconstituted or formed anew only in the 1990s, after the beginning of the political transformations. Under the old repressive system, only the voluntary fire brigades had been able to continue working—even through the 1950s. The best-known local organization is the Kapolcs Nature Preservation and Cultural Society, which organizes the summer cultural events. An old, now revitalized organization is the Farmers Circle. There is also a soccer club. These organizations have expressed their intention to work together with their local government to develop the basin in a sustainable, nature-friendly manner.

The basin's annual cultural festival started in one village in 1987. By 1998, it had been extended to all five villages. Some 120,000 visitors occupy the area for the nine hottest days of July to hear concerts, watch theatrical performances, and attend exhibitions. The community places, including the library, the church, the cultural house, the soccer field, and—God forbid!—the pub, are open to multiple uses.

Developing a Vision for the Community

Planning for the Dörögd Basin's development program was initiated early on by participants of the community college, who issued a "letter of commitment" inviting further cooperative community action. Subsequently, following an inventory of the physical resources and the natural and cultural values of the local area, a strategy of organic development has been agreed upon. The collaborative planning effort was based on the carefully organized public participation process but also included experts who, working from the recommendations on both regional and village levels, summarized the objectives.

The community college served as the vehicle by which questions concerning the future of the basin were put to the local people. Discussions and lectures revealed the advantages and disadvantages of life in the Dörögd Basin. Questions were asked about prospects defined from both the individual and the community perspective, drawing attention also to possible visions of a desirable future. The two basic questions raised for the adults in the community were:

1. Where do you think our strengths lie for the next ten years?
2. What are the good things that give us confidence regarding the future and could become a "pull" to the area?

The answers stressed the following strengths, which in turn suggested which values ought to be protected:

1. The landscape,
2. The quiet, the tranquility,
3. Clean air,
4. The advantages of the location of the basin (close to Lake Balaton, yet removed from the noisy tourist hordes),
5. The legal protection of certain natural and cultural assets and values,
6. The diligence of local residents,
7. Time-honored skills and professions,
8. Herb gathering,
9. The hospitality of the folk,
10. The young people,
11. Good local management,
and others.

Ideas for aims that could realistically be achieved were also collected:

1. Making pictogram information boards (traffic signs)
2. Maintaining roads,
3. Creating jobs,
4. Increasing the number of organic kitchen gardens,
5. Engaging retired master craftsmen,
6. Increasing hospitable family house lettings,
7. Strengthening the cooperation of NGOs,
8. Increasing the number of hours spent working for the community,
and so forth.

Children entered a picture drawing competition with the theme "The Future of the Dörögd Basin." The majority of the children's drawings reflected the beauty and peace of this small region. Only a few of the artists stepped outside of the prevailing ethos of rustic tranquility; generally, the children expressed positive feelings in their pictures. Despite the fact that many adults said otherwise, many children considered country farming practices natural.

A Program of Organic Development to Strengthen Local Sustainability

Based on the "letter of commitment" signed by the main stakeholders in 1993 on the one hand, and on the values expressed in the "spontaneous future vision" questioning process on the other, a comprehensive strategy of organic development was elaborated.

Conservation

The objective of the organic development program was to initiate effective communication among different stakeholders of the community in order to achieve environmental education in the broadest sense as well as to strengthen local capacities for nature conservation. The Dörögd Basin watershed seemed to be a suitable arena for launching such a program. An obvious starting point was the inventory of natural values, which was published and disseminated to the members of the community. All active conservation projects subsequently were grounded in the increased awareness engendered by this catalogue.

Organic Production

Almost from the beginning the main—and most difficult—objective of the program was to create an environmentally sensitive economic base of existence for the inhabitants. Agriculture based on organic methods would most closely satisfy the ideal. It would be naive to think, however, that lectures, courses, diplomas, and the few available contracts alone could change the existing profile of this means of livelihood in the short run. However, those who undertook to convert to organic methods could gradually remodel their practices by simple, "down-to-earth" steps, for example first reducing and then eliminating the use of pesticides.

Community Development

Environmental protection and community development are instrumental to each other. For seven years, the mayors of the villages were invited bimonthly to discuss issues before decisions were taken involving conservation issues. While doing so, democratic procedures were used and collaborative planning was maintained.

Scenarios for Autonomous Infrastructure and Sustainable Tourism: Expert Recommendations

The possibility of changing the structure of energy consumption and energy policy has been on the agenda of increasingly higher levels of government since the 1970s. From the beginning of the transition, sustainable use of renewable energy resources has been set as a special priority within these concepts. Accordingly, the Dörögd development program sought to set up a working idea for solving the energy requirements of the basin that would be compatible with the traditional landscape and with the agricultural and human values of the basin, using long-term sustainable resources. The feasibility studies completed here regarding the restoration of a sustainable landscape, the further use of renewable energies, and decentralized, alternative sewage treatment have been published as model recommendations at the national level.

Education

Within the Dörögd Basin, several generations side by side have been training to recognize the natural values of the area and ways to protect them. Courses were held on organic gardening, organic agriculture, and organic production of herbs. Also, courses for future hosts and hostesses young and old paved the way for sustainable tourism in the basin. In the school garden, local students learn how to work with herbs while using modern, solid technology, i.e. solar cells for drying. Educational opportunities such as these are designed to keep local youth in the small region. In the last few years, systematic education has been offered on landscapes and their protection, on sustainable forestry, on traditional life, etc., in the form of field studies for students and postgraduate teachers' courses that included visits to the education center of the IEC in Budapest.

Lessons Learned: Collaborative Planning and Cross-Generational Education as Instruments of Local Sustainable Development

Analysis of recent experience in the Dörögd Basin shows tangible results of certain actions and approaches that have been taken there. Moreover, the effects have been complex, with positive interreactions between the environmental, economic, and social spheres. In particular, the following actions, grouped under broad categories, may be considered for general application:

Active conservation:
1. "Green corridors": planting an eleven-hectare native forest, planting an alley of linden trees five kilometers long, etc.,
2. Repairing and restoring natural stream banks,

3. Re-creating a mosaic part of the National Park,
4. Re-creating native orchards with technical and financial assistance,

Organic production:
5. Advising farmers and schools on organic gardening and agriculture and small scale food processing,
6. Growing organic herbs; using community-owned dehydrators to process them; initiating the cooperative "HERBIO,"

Collaborative planning and community development:
7. Strategic planning for complex organic development of the basin,
8. Feasibility studies on landscape protection, the use of renewable energies, and alternative sewage treatment,
9. Promotion of a concept of sustainable tourism,
10. Cooperative implementation of the strategy involving, for example, the local governments and youth,

Education and training for the local adult and youth population as well as for visitors:
11. Enhancing local awareness of the local natural heritage,
12. Developing community college courses and field-trips,
13. Offering environmental education for the local students and assistance in the school garden,
14. Organizing forums for farmers, entrepreneurs, civic organizations, mayors,
15. Establishing a field study center for landscape protection that invites students from outside the local area,
16. Publishing books and CD-ROMs; posting information on the Internet.

Criteria to be considered in evaluating such actions or approaches should include (a) the protection of both biological and cultural diversity, (b) the extension of community autonomy, and (c) the restoration of local identity.

SELECT BIBLIOGRAPHY

Abbey, Edward. *Down the River.* New York: E.P. Dutton, 1982.
Altvater, E. "Theoretical Deliberations on Time and Space in Post-socialist Transformation." *Regional Studies* 32 (1998): 591–605.
Amato, Anthony J. "In the Wild Mountains: Idiom, Economy, and Ideology among the Hutsuls of Ukraine, 1849–1939" (Ph.D. dissertation, Indiana University, 1998).
Andonova, Liliana B. *Transnational Politics of the Environment: The European Union and Environmental Policy in Central and Eastern Europe.* Cambridge, MA: MIT Press, 2004.
Auer, Matthew, ed. *Restoring Cursed Earth: Appraising Environmental Policy Reforms in Eastern Europe and Russia.* Lanham, MD: Rowman and Littlefield, 2004.
Augustyn, Maciej, and Kajetan Perzanowski. *Selected Ecological Problems of [the] Polish-Ukrainian Carpathians.* Bieszczady, Poland: Polish Academy of Sciences, 1997.
Baker, Susan, and Petr Jehlicka, eds. *Dilemmas of Transition: The Environment, Democracy and Economic Reform in East Central Europe.* London: Frank Cass, 1998.
Balcerowicz, Leszek. "The Polish Way to the Market Economy." In *United States Relations with Central and Eastern Europe,* ed. Dick Clark, 7–14. Queenstown, MD: Aspen Institute, 1993.
Becattini, G. "The Marshallian ID as a socio-economic notion." In *Industrial Districts and Inter-firm Co-operation in Italy,* ed. F. Pyke, G. Becattini, W. Sengenberger, 37–51. Geneva: International Institute for Labour Studies, 1991.
Beckmann, Andreas, JoAnn Carmin, and Barbara Hicks. "Catalysts for Sustainability: NGOs and Regional Development Initiatives in the Czech Republic." In *International Experiences on Sustainability,* ed. Walter Leal Filho, 159–77. Bern: Peter Lang Scientific Publishing, 2002.
Beckmann, Andreas, and Henrik Dissing. "EU Enlargement and Sustainable Rural Development in Central and Eastern Europe." *Environmental Politics* 13, no. 1 (2004): 135–52.
Bergman, E. M. "Industrial Trade Clusters in Action: Seeing Regional Economies Whole." In *Clusters and Regional Specialization,* ed. M. Steiner, 92–110. London: Pion, 1998.
Bergman, E. M., and E. J. Feser. "Innovation System Effects on Technological Adoption in a Regional Value Chain." *European Planning Studies* 9, no. 5 (2001): 629–48.
Bergman, E. M., and Harvey Goldstein. "Urban Innovation and Technological Advance in the Research Triangle Region." In *Sustainability of Urban Systems: A Cross-National Evolutionary Analysis of Urban Innovation,* ed. P. Nijkamp. London: Gower, 1990.

Bilsen, V., and J. Konings. "Job Creation, Job Destruction, and Growth of Newly Established, Privatized and State-Owned Enterprises in Transition Economies: Survey Evidence from Bulgaria, Hungary and Romania." *Journal of Comparative Economics* 26 (1998): 429–45.
Bochniarz, Zbigniew. "Environmental Concerns in Central and Eastern Europe: Challenges and Solutions." In *United States Relations with Central and Eastern Europe,* ed. Dick Clark, 15–25. Queenstown, MD: Aspen Institute, 1993.
Bochniarz, Zbigniew, and David Toft. "Free Trade and the Environment in Central Europe." *European Environment* 5 (1995): 52–7.
Bochniarz, Zbigniew, Wladyslaw Jermakowicz, and David Toft. "Strategic Foreign Investors and the Environment in Central and Eastern Europe." In *Innovation, Technology and Information Management for Global Development and Competitiveness,* ed. Erdener Kaynak and Tunc Erem. Hummelstown, PA: International Management Development Association, 1995.
Bochniarz, Zbigniew, and R. S. Bolan. *Institutional Design for Financing Sustainable Development: Lessons Learned from Poland.* Minneapolis: University of Minnesota, 2000.
Boschma, R. A., and J. G. Lambooy. "Evolutionary Economics and Economic Geography." *Journal of Evolutionary Economics,* 9 (1999): 411–29.
Bosquet, B. "Environmental Tax Reform: Does It Work? A Survey of the Empirical Evidence." *Ecological Economics* 34, no. 1 (2000): 19–32.
Bovenberg, A. L., and L. H. Goulder. "Costs of Environmentally Motivated Taxes in the Presence of other Taxes: General Equilibrium Analyses." *National Tax Journal* 50, no. 1 (1997): 59–87.
Brenner, N. "Globalisation as Reterritorialisation: The Re-scaling of Urban Governance in the European Union." *Urban Studies* 36 (1999): 431–51.
Brown, L. R. *Building a Sustainable Society.* New York: W. W. Norton, 1981.
Cabus, P. "The Meaning of Local in a Global Economy: The 'Region's Advocacy of Local Interests' as a Necessary Component of Current Global/Local Theories." *European Planning Studies* 9 (2001): 1011–29.
Carius, Alexander. "Challenges for Governance in a Pan-European Environment: Transborder Cooperation and Institutional Coordination." In *EU Enlargement and Environmental Quality: Central and Eastern Europe & Beyond,* ed. S. Crisen and J. Carmin. Washington, D.C.: Woodrow Wilson International Center for Scholars, 2002.
Carmin, JoAnn, and Barbara Hicks. "International Triggering Events, Transnational Networks, and the Development of the Czech and Polish Environmental Movements." *Mobilization* 7, no. 3 (2002): 305–24.
Carmin, JoAnn, and Stacy D. VanDeveer. *EU Enlargement and the Environment: Institutional Change and Environmental Policy in Central and Eastern Europe.* London: Routledge, 2005.
Carter, F. W., and David Turnock. *Environmental Problems in Eastern Europe.* London: Routledge, 1996.
Clark, G. L., M. P. Feldman, and M. S. Gertler, eds. *The Oxford Handbook of Economic Geography.* Oxford: Oxford University Press, 2000.
Clark, T. N. "Old and New Paradigms for Urban Research: Globalization and the Fiscal Austerity and Urban Innovation Project." *Urban Affairs Review* 36 (2000): 3–45.
Claval, P. "Reflexions sur la centralité." *Cahiers de Géographie du Québec* 44 (2000): 285–301.
Cole, John W., and Eric R. Wolf. *The Hidden Frontier: Ecology and Ethnicity in an Alpine Valley.* New York: Academic Press, 1974.

Crisen, Sabina, and JoAnn Carmin, eds. *EU Enlargement and Environmental Quality: Central and Eastern Europe & Beyond.* Washington, D.C.: Woodrow Wilson International Center for Scholars, 2002.
Dabrowski, Marek, Stanislaw Gomulka, and Jacek Rostowski. "Whence Reform? A Critique of the Stiglitz Perspective." *Journal of Policy Reform* 4 (2001): 291–324.
den Hertog, P., E. M. Bergman, and D. Charles, eds. *Innovative Clusters: Drivers of National Innovation Systems.* Paris: OECD Proceedings, 2001.
Dingsdale, A. "Central Places and Planning in Hungary." *Geography Review* 6 (1993): 12–16.
Dubos, René. *The Wooing of the Earth.* New York: Charles Scribner's Sons, 1980.
Economic Commission for Europe. *Economic Survey for Europe.* New York: United Nations Publications, 2001.
Ertsey, Attila, Judit Vásárhelyi, and Péter Medgyasszay, eds. *Autonóm Kisrégió. Országos ajánlás.* Budapest: Független Ökológiai Központ, 1999.
Fagin, A. "The Development of Civil Society in the Czech Republic: The Environmental Sector as a Measure of Associational Activity." *Journal of European Area Studies* 7 (1999): 91–108.
Feser, E. "Old and New Theories of Industrial Clusters." In *Clusters and Regional Specialization,* ed. M. Steiner, 18–40. London: Pion, 1998.
Feshbach, Murray, and Alfred Friendly. *Ecocide in the USSR.* Foreword by Lester Brown. New York: Basic Books, 1991.
Francis, Patrick, ed. *National Environmental Protection Funds: Case Studies of Bulgaria, the Czech Republic, Hungary, Poland and the Slovak Republic.* Budapest: Regional Environmental Center, 1994.
Fritz, O. M., H. Mahringer, and M. T. Valderrama. "A Risk-oriented Analysis of Regional Clusters." In *Clusters and Regional Specialization,* ed. M. Steiner, 181–91. London: Pion, 1998.
Garvey, Tom. "EU Enlargement: Is It Sustainable?" In *EU Enlargement and Environmental Quality: Central and Eastern Europe & Beyond,* ed. S. Crisen and J. Carmin, 53–62. Washington, D.C.: Woodrow Wilson International Center for Scholars, 2002.
Glaeser, E. L., H. D. Kallal, J. A. Sheinkman, and A. Scheifler. "Growth in Cities." *Journal of Political Economy* 100 (1992): 1126–52.
Göttke-Krogmann, Ulrich. "Huzulen-Vergangenheit und Gegenwart." *Österreichische Osthefte* 42, no. 3–4 (2000): 109–38.
Greer-Wootten, B. "Sustainability and Scale." In *Regional Prosperity and Sustainability,* ed. P. Hlavinková and J. Munzar, 75–85. Brno: Regiograph for Geokonfin, 1999.
———. "Towards a Geography of Sustainable Development and/or Sustainability?" *Moravian Geographical Reports* 12 (2004): 2–9.
Grindle, Merilee S. *Getting Good Government: Capacity Building in the Public Sector of Developing Countries.* Cambridge, MA: Harvard University Press, 1997.
Grossman, G. M., and A. B. Krueger. "Economic Growth and the Environment." *The Quarterly Journal of Economics* (May 1995): 353–77.
Gudowski, Janusz. "Market Oriented Transformation of the Rural Society in Mountain Zones: Example of Carpathian Villages in Western Ukraine." In *Transforming Rural Sector to the Requirements of Market Economy: Examples from Turkey, Poland and Ukraine,* ed. Gülcan Eraktan and Janusz Gudowski. Warsaw: Dialog, 1997.
Guminska, Maria, and Andrzej Delorme, eds. *Kleska Ekologiczna Krakowa.* Cracow: PKE Krakow, 1990.
Gürsan-Salzmann, Ayşe. "Shepherds of Transylvania." *Natural History* (July 1984): 42–52.

Gutner, Tamar L. *Banking on the Environment: Multilateral Development Banks and Their Environmental Performance in Central and Eastern Europe.* Cambridge, MA: The MIT Press, 2002.

Hallstrom, Lars K. "Eurocratising Enlargement? EU Elites and NGO Participation in European Environmental Policy." *Environmental Politics* 13, no. 1 (2004): 175–93.

Hardi, P. *Impediments on Environmental Policy-Making and Implementation in Central and Eastern Europe: Tabula Rasa vs. Legacy of the Past.* Policy Papers in International Affairs 40. Berkeley: Institute of International Studies, University of California, 1992.

Henderson, V., A. Kuncoro, and M. Turner. "Industrial Development in Cities." *Journal of Political Economy* 103, no 5. (1995): 1067–90.

Hicks, Barbara. "Setting Agendas and Shaping Activism: EU Influence on Central European Environmental Movements." *Environmental Politics* 13, no. 1 (2004): 216–33.

Hinderink, J., and M. Titus. "Small Towns and Regional Development: Major Findings and Policy Implications from Comparative Research." *Urban Studies* 39 (2002): 379–91.

Holmén, Hans. "The Unsustainability of Development." *International Journal of Economic Development* 3, no. 1 (2001): 1–26.

Holtz-Eakin, D., and T. M. Selden. "Stoking the Fires? CO_2 Emissions and Economic Growth." *Journal of Public Economics* 57 (1995): 85–101.

Holubets', Mykhailo, ed. *Ukrainskie Karpaty: Priroda.* Kiev: Naukova Dumka, 1988.

———. *Antropohenni zminy bioheotsenotychnoho pokryvu v Karpats'komu rehioni.* Kiev: Naukova Dumka, 1994.

Holzinger, Katerine, and Peter Knoepfel. *Environmental Policy in a European Union of Variable Geometry? The Challenge of the Next Enlargement.* Basel: Helbing and Lichtenhahn, 2000.

Homeyer, Ingmar von. "Differential Effects of Enlargement on EU Environmental Governance." *Environmental Politics* 13, no. 1 (2004): 52–76.

Hopkins, Alan, ed. *Grass: Its Production and Utilization.* London: Blackwell, 2000.

Hoshko, H. Iu., and R. F. Kyrchiv. *Hutsul'shchyna: istoryko-etnohrafichne doslidzhennia.* Kiev: Naukova Dumka, 1987.

Hryniuk, Stella. *Peasants with Promise: Ukrainians in Southeastern Galicia 1880–1900.* Edmonton: Canadian Institute of Ukrainian Studies, 1991.

Ingerson, Alice E. "Testing the Nature/Culture Dichotomy in Practice." In *Historical Ecology: Cultural Knowledge and Changing Landscapes,* ed. Carole L. Crumley, 43–66. Santa Fe: School of American Research Press, 1994.

Jancar-Webster, Barbara. "Environmental Movement and Social Change in the Transition Countries." In *Dilemmas of Transition: The Environment, Democracy and Economic Reform in East Central Europe,* ed. Susan Baker and Petr Jehlicka, 69–92. London: Frank Cass, 1998.

Jehlicka, Petr, and Andrew Tickle. "Environmental Implications of Eastern Enlargement: The End of EU Progressive Environmental Policy?" *Environmental Politics* 13, no. 1 (2002): 79–95.

Kaczmarek, Z., J. J. Napiorkowski, and D. Jurak. *Impact of Climate Change on Water Resources in Poland.* Institute of Geophysics, Polish Academy of Sciences no. 295, Series E-1. Warsaw: The Institute of Geophysics, 1997.

Kallabová, E. "Changes of Hierarchical Positions of Small Moravian Towns on the Example of Education." In *Nature and Society in Regional Context,* ed. P. Hlavinková and J. Munzar, 74–82. Brno: Regiograph for Geokonfin, 2001.

Klaassen, Ger, and Mark Smith. *Financing in Environmental Change in Central and Eastern Europe: An Assessment of International Support.* Laxenburg, Austria: International Institute for Applied Systems Analysis, 1995.

Kolk, Ans, and Ewout van der Weij. "Financing Environmental Policy in East Central Europe." In *Dilemmas of Transition: The Environment, Democracy and Economic Reform in East Central Europe,* ed. Susan Baker and Petr Jehlicka, 53–68. London: Frank Cass, 1998.

Komendar, V. I. *Likars'ki roslyny Karpat.* Uzhhorod: Karpaty, 1971.

Komlos, John. *The Habsburg Monarchy as a Customs Union: Economic Development in Austria-Hungary in the Nineteenth Century.* Princeton: Princeton University Press, 1982.

Kozak, Ihor. "Antropohenna transformatsiia roslynnoho pokryvu hirs'koï chastyny baseinu r. Prut." *Ukraïns'kyi botanychnyi zhurnal* 47, no. 2 (1990): 59–64.

Kozak, Ihor, and Mykhailo Holubets'. "Lisovyi bioheotsenotychnyi kompleks verkiv'ia Pruta." In *Antropohenni zminy bioheotsenotychnoho pokryvu v Karpats'komu rehioni,* ed. Mykhailo Holubets', 35–45. Kiev: Naukova Dumka, 1994.

Kozak, Ihor, and Maciej Augustyn. "The Trends of Anthropogenic Pressure in [the] Polish and Ukrainian Carpathians." In *Selected Ecological Problems of [the] Polish-Ukrainian Carpathians,* ed. Kajetan Perzanowski and Maciej Augustyn, 15–21. Bieszczady, Poland: Polish Academy of Sciences, 1997.

Kozak, Ihor, and Vladimir Menshutkin. "An Investigation of a Mixed Beech Forest Dynamics in [the] Ukrainian Carpathians Using a Computer Model." In *Selected Ecological Problems of [the] Polish-Ukrainian Carpathians,* ed. Kajetan Perzanowski and Maciej Augustyn, 23–8. Bieszczady, Poland: Polish Academy of Sciences, 1997.

Kramer, John M. "EU Enlargement and the Environment: Six Challenges." *Environmental Politics* 13, no. 1 (2004): 290–311.

Kruger, Christine, and Alexander Carius. *Environmental Policy and Law in Romania: Towards EU Accession.* Berlin: Ecologic, 2001.

Kurek, Włodzimierz. "Agriculture Versus Tourism in Rural Areas of the Polish Carpathians." *GeoJournal* 38, no. 2 (1996): 191–6.

Leopold, Aldo. *A Sand County Almanac with Essays on Conservation from Round River.* New York: Ballantine Books, 1980.

Levy, Barry S., ed. *Air Pollution in Central and Eastern Europe: Health and Public Policy.* Boston: Management Sciences for Health, 1991.

Lietzman, K. M., and G. D. Vest. *Environment and Security in an International Context.* Bonn and Washington, D.C.: North Atlantic Treaty Organization, 1999.

Maier, G., and E. M. Bergman. "Conjoint Analysis of Transport Options in Austrian Regions and Industrial Clusters." In *Freight Transport Demand and Stated Preference Experiments,* ed. R. Danielis. Milan: FrancoAngeli, 2002.

Mandybura, Mar'ian Danylovych. *Polonyns'ke hospodarstvo Hutsul'shchyny druhoï polovyny XIX-30-x rokiv XX st.* Kiev: Naukova Dumka, 1978.

Markowska, Agnieszka, and Tomasz Zylicz. "Costing on International Public Good: The Case of the Baltic Sea." *Ecological Economics* 30 (1999): 301–16.

Maskell, P., and A. Malmberg. "Localized Learning and Industrial Competitiveness." *Cambridge Journal of Economics* 23 (1999): 167–85.

Mayer, H. J., and M. R. Greenberg. "Coming Back From Economic Despair: Case Studies of Small- and Medium-Size American Cities." *Economic Development Quarterly* 15, no. 3 (2001): 203–16.

McGinnis, M. V. *Bioregionalism.* London: Routledge, 1998.

Möller, L., *Umweltpolitik im Transformationsprozess: Die Beispiele Polen und Tschechische Republik.* Marburg: Metropolis, 2002.

Molnar, D., A. Morgan, and D. V. J. Bell. *Defining Sustainability, Sustainable Development and Sustainable Communities.* Toronto: Sustainable Toronto Project, 2001.

Munduch, E. M., and A. Spiegler. *Kleinstädte: Motoren in ländlichen Raum.* Landtechnischen Schriftenreihe, no. 214. Vienna: Österreichisches Kuratorium für Landtechnik und Landentwicklung, 1998.

Námer, J., M. Drtil, I. Bodík, and M. Hutňan. "Wastewater treatment in [the] Slovak Republic." *Polish Journal of Environmental Studies* 6, no. 2 (1997): 39–45.

National Research Council, Committee on the Science of Climate Change. *Climate Change Science: An Analysis of Some Key Questions.* Washington, D.C.: National Academy Press, 2001.

Niedermayer, M. *Kleinstadtentwicklung.* Würzburg: Geographisches Institut der Universität, 2000.

Olson, M. *The Rise and Decline of Nations: Economic Growth, Stagflation and Social Rigidities.* New Haven: Yale University Press, 1982.

O'Riordan, T., ed. *Ecotaxation.* London: Earthscan, 1997.

Pavlinek, Petr, and John Pickles. *Environmental Transitions: Transformation and Ecological Defence in Central and Eastern Europe.* London: Routledge, 2000.

———. "Environmental Pasts/Environmental Futures in Post-Socialist Europe." *Environmental Politics* 13, no. 1 (2004): 237–65.

Pavliuk, Stepan P. *Narodna ahrotekhnika ukraïntsiv Karpat druhoï polovyni XIX-pochatku st.* Kiev: Naukova Dumka, 1986.

Pearce, David W. *Blueprint 3: Measuring Sustainable Development.* London: Earthscan, 1993.

Pearce, David W., and Jeremy J. Warford. *World Without End: Economics, Environment and Sustainable Development.* New York: Oxford University Press, 1993.

Peneder, M. "Dynamics of Initial Cluster Formation: The Case of Multimedia and Cultural Content." In *Innovative Clusters: Drivers of National Innovation Systems,* ed. P. den Hertog, E. M. Bergman, and D. Charles, 303–12. Paris: OECD Proceedings, 2001.

Peterson, D. J. *Troubled Lands: The Legacy of Soviet Environmental Destruction.* Boulder: Westview Press, 1993.

Porter, M. E. *The Competitive Advantages of Nations.* London: McMillan, 1990.

———. "Competitive Advantage, Agglomeration Economies and Regional Policy." *International Regional Science Review* 19, no. 1 (1996): 85–94.

Potůček, M., et al. *Průvodce krajinou priorit pro Českou republicku* (A guide to a priorities landscape for the Czech Republic). Prague: Centrum pro sociální a ekonomické strategie, Universita Karlova v Praze, Fakulta sociálních věd, 2002.

Pryde, Philip R. *Conservation in the Soviet Union.* Cambridge: Cambridge University Press, 1972.

———. *Environmental Management in the Soviet Union.* Cambridge: Cambridge University Press, 1992.

Pyne, Stephen J. *World Fire: The Culture of Fire on Earth.* New York: Henry Holt and Company, 1995.

Quigley, Kevin F. F. "Lofty Goals, Modest Results: Assisting Civil Society in Eastern Europe." In *Funding Virtue: Civil Society Aid and Democracy Promotion,* ed. Thomas Carothers and Marina Ottaway, 191–216. New York: The Carnegie Endowment, 2000.

Roelandt, T. J. A., and P. den Hertog, eds. *Boosting Innovation: The Cluster Approach.* Paris: OECD Proceedings, 1999.

Rosenfeld, S. A. "Bringing Business Clusters into the Mainstream of Economic Development." *European Planning Studies* 5, no. 1 (1997): 3–23.

Rudolph, Richard. "The East European Peasant Household and the Beginnings of Industry: East Galicia, 1786–1914." In *Ukrainian Economic History,* ed. I. S. Koropeckyj, 339–82. Cambridge, MA: Harvard University Press, 1991.

Russell, Emily W. B. *People and the Land Through Time: Linking Ecology and History.* New Haven: Yale University Press, 1997.
Sagar, Ambuj. "Capacity Development for the Environment: A View from the South, A View from the North." *Annual Review of Energy and the Environment* 25 (2000): 377–439.
Sandberg, Mikael. *Green Post-Communism: Environmental Aid, Polish Innovation and Evolutionary Political Economics.* London: Routledge, 1999.
Schlegelmilch, K., ed. *Green Budget Reform in European Countries.* Berlin: Springer, 1999.
Schreurs, Miranda A., and Elizabeth Economy. *The Internationalization of Environmental Protection.* Cambridge: Cambridge University Press, 1997.
Senkiv, Ivan. *Die Hirtenkultur die Huzulen. Eine volkskundliche Studie.* Marburg/Lahn: J. G. Herder Institut, 1981.
Somlyódy, L. "Quo vadis Water Quality Management in Central and Eastern Europe." *Water Science Technology* 30, no. 5 (1994): 1–14.
Stanners, David, and Philippe Bourddeau, eds. *Europe's Environment: The Dobris Assessment.* Copenhagen: European Environment Agency, 1995.
Stern, David I., Michael S. Common, and Edward B. Barbier. "Economic Growth and Environmental Degradation: The Environmental Kuznets Curve and Sustainable Development." *World Development* 24 (1996): 1151–60.
Stiglitz, Joseph, and David Ellerman. "Not Poles Apart: 'Whither Reform?' and 'Whence Reform?'" *Journal of Policy Reform* 4 (2001): 325–39.
Tarras-Wahlberga, N. H., et al. "Environmental impacts and metal exposure of aquatic ecosystems in rivers contaminated by small scale gold mining: The Puyango River basin, southern Ecuador." *The Science of the Total Environment* 278 (2001): 239–61.
Tichy, G. "Clusters: Less Dispensable and More Risky than Ever." In *Clusters and Regional Specialization,* ed. M. Steiner, 211–25. London: Pion, 1998.
Toman, Michael. *Pollution Abatement Strategies in Central and Eastern Europe.* Washington, D.C.: Resources for the Future, 1994.
Třebický, V. "Ekologická stopa" (The ecological footprint). In *Unese Země civilizaci?* (Will the Earth kidnap civilization?). Prague: Ministry of the Environment, 2000.
VanDeveer, Stacy D. "Normative Force: The State, Transnational Norms and International Environmental Regimes." Ph.D. dissertation, University of Maryland, College Park, MD, 1997.
———. "Europeanizing Central Europe: Capacity, Surprises, Lessons and Challenges." In *EU Enlargement and Environmental Quality: Central and Eastern Europe & Beyond,* ed. S. Crisen and J. Carmin, 114–22. Washington, D.C.: Woodrow Wilson International Center for Scholars, 2002.
VanDeveer, Stacy D., and Geoffrey D. Dabelko. *Protecting Regional Seas: Developing Capacity and Fostering Environmental Cooperation in Europe.* Washington, D.C.: Woodrow Wilson International Center for Scholars, 2000.
VanDeveer, Stacy D., and Geoffrey D. Dabelko. "It's Capacity, Stupid: National Implementation and International Assistance." *Global Environmental Politics* 1, no. 2 (2001): 18–29.
VanDeveer, Stacy D. and Ambuj Sagar. "Capacity Building for the Environment: North and South." In *Furthering Consensus: Meeting the Challenges of Sustainable Development Beyond 2002,* ed. E. Corell, A. Churie Kallhauge, and G. Sjöstedt. London: Greenleaf (forthcoming).
Vaňo, B. *Prognóza vývoja rómskeho obyvateľstva v SR do roku 2025.* Bratislava: Výskumné demografické centrum, 2002.
Vásárhelyi, Judit. "Reports on the Relationship between Environmental NGOs and Governments in Central and Eastern Europe—the NGO Perspective." In *New Horizons?*

Possibilities for Cooperation between Environmental NGOs and Governments in Central and Eastern Europe, ed. P. Hardi, Alexander Juras, and Magda Tóth Nagy, 139–45. Budapest: Regional Environmental Center for Central and Eastern Europe, 1993.

———. "Local Initiatives for Collaborative Environmental Planning in Urban and Rural Communities." In *Hungary: Towards Strategy Planning for Sustainable Development: National Information to the United Nations Commission on Sustainable Development,* 41–3. Budapest: Fenntartható Fejlődés Bizottság, 1996.

———. "Strengthening Local Capacities for Water Conservation in a Small Catchment Area: Public Participation in the Dörögd Basin, Balaton Upperland." In *Participatory Processes in Water Management: Proceedings of the Satellite Conference to the World Conference on Science,* ed. J. Gayer, 119–128. Budapest, Hungary, 28–30 June 1999. Paris: UNESCO, 1999.

———. "Közösségi tervezés." In *A szag nyomában. Környezeti konfliktusok és a helyi társadalom,* ed. Lányi András, 286–96. (Budapest: Osiris, 2001.

———. "Közösségek identitása, autonómiája." In *Vissza vagy hova? Útkeresés a fenntarthatóság felé Magyarországon,* ed. Tamás Pálvölgyi, Csaba Nemes, and Zsuzsanna Tamás, 81–91. Budapest: K. Tertia, 2002.

———, ed. "Initiatives of Citizens Groups, Programs of Local and Regional Governments." In part 4 of *Hungary: Strategy, Plans, Initiatives and Actions for Sustainable Development,* 92–103. Budapest: Fenntartható Fejlődés Bizottság, 1997.

Vásárhelyi, J., and S. McIlwane. "Collaborative Environmental Planning on the Community Level: An Adaptation of the Comparative Risk Model in Two Hungarian Communities." In *Proceedings, International Workshop in Public Participation in Environmental Decisions: A Challenge for Central and Eastern Europe.* Budapest: Regional Environmental Center for Central and Eastern Europe, 1992.

Vásárhelyi, Judit, and Anna Vári. "A Dörögdi-medence organikus fejlesztési programjának tapasztalatai." *Társadalomkutatás* 1–2 (1999): 60–75.

———. "Public Involvement in Local Environmental Planning: The Case of Sátoraljaújhely." In *Public Participation in Environmental Decisions,* ed. A. Vári and J. Caddy, 110–26. Budapest: Akad. Kiad., 1999.

Vašečka, M. "Rómovia." In *Slovensko, Súhrnná správa o stave spoločnosti,* ed. M. Kollár and G. Mesežnikov. Bratislava: Inštitút pre verejné otázky, 2000.

Vincenz, Stanisław de. *On the High Uplands: Sagas, Songs, Tales, and Legends of the Carpathians.* Trans. H. C. Stevens. London: Hutchinson, 1955.

Vogel, W. R. "Water Monitoring in the Light of the EU Water Framework Directive— The Austrian Approach." In *Proceedings: International Conference on EU Water Management Framework Directive and Danubian Countries: Bratislava, 21–23 June 1999,* ed. Eva Pálmaiová, 234–9. Bratislava: Stimul, 1999.

———. "The Austrian Water Monitoring System and Its Integration into the Water Management Concept." In *Environmental Toxicology Assessment,* ed. Mervyn Richardson, 227–46. London: Taylor and Francis, 1995.

Vogel, W. R., J. Grath, G. Winkler, and A. Chovanec. "The Austrian Water Monitoring System—Information for Different Levels of the Decision Making Processes." In *Proceedings: Monitoring Tailor Made,* vol. 2, 147–53. Nunspeet, 1996.

Wackernagel, Mathis, and William Rees. *Our Ecological Footprint: Reducing Human Impact on the Earth.* Philadelphia: New Society Publishers, 1996.

Wedel, Janine R. *Collision and Collusion: The Strange Case of Western Aid to Eastern Europe 1989–1998.* New York: St. Martin's Press, 1998.

Wiener, Douglas R. "The Changing Face of Soviet Conservation." In *The Ends of the Earth: Perspectives on Modern Environmental History*, ed. Donald Worster, 252–73. Cambridge: Cambridge University Press, 1988.

Wódz, Jacek, and Kazimierza Wódz. "Environmental Sociology in Poland and the Ecological Consciousness of the Polish People." In *Environment and Society in Eastern Europe*, ed. Andrew Tickle and Ian Welsh, 97–113. New York: Longman, 1998.

INDEX

A

Aarhus Convention. *See* United Nations Economic Commission for Europe (UNECE)
accession. *See* European Union (EU)
acquis communautaire. *See* European Union (EU)
air quality, 35, 38–40, 43, 46, 52, 232
Albania, 12, 22–23, 33, 168
Armenia, 22–23, 33
Aurul, S.A., 129, 131–132, 134, 138
Austria, 1–3, 5, 7, 14, 59, 67–68, 73, 76–77, 84, 124–126, 143–149, 169, 172–173, 175, 178, 183
Autonomous infrastructure, 244. *See also* Dörögd Basin
Azerbaijan, 22–23, 33

B

Baia Borsa, 133
Baia Mare, 7, 131–142
Balaton, Lake, 14, 236–238, 242
Baltic Sea, 2, 36, 158
Baltic states, 12, 21, 29–30, 36, 39, 42, 46
Belarus, 8, 12, 22–23, 33, 150, 152–154, 156, 164–165, 224
"Black Triangle", 2
Bosnia and Herzegovina, 169–170
Brno, 177, 224, 228
Bulgaria, 1, 2, 4, 20, 22–23, 33, 50, 53, 57, 147, 169, 173, 179

C

capacity building, 4, 46–49, 52, 55, 160
capacity development, 4, 45–58
Capacity development for the environment (CDE), 46, 52, 55
Carpathians, 8, 183–202. *See also* Galicia; and Hutsul region
Carpathian basin, 235
Central Place Theory, 219–220
"central settlement system", 224
Cheremosh River, 189
civil society, 21, 45–47, 49, 51, 54, 209, 217–218, 229
climate change, 3, 5–6, 11, 13–15, 40, 81, 83–84, 87, 96–97, 126
clusters (and regions), 5, 59–77
coal, 2, 12, 14, 46, 60, 82–83, 87, 94, 103–104, 106–107, 109–110, 113, 115–117, 226, 239
commons, 197, 235
communism, 1–3, 5, 11, 45–46, 49, 51, 54, 81, 122, 151, 167–168, 170, 173, 175, 197, 232, 240
computable general equilibrium modeling (CGE), 99, 102, 105, 111, 112. *See also* general equilibrium modeling
Croatia, 22–23, 33, 83, 98, 147, 169
cyanide, 2, 7, 129–137, 139, 141–142
Cyprus, 50
Czechoslovakia, 1–3, 12, 46, 123, 203, 218
Czech Republic, 5–7, 14, 20–23, 33, 41, 46, 48, 50, 52, 81–98, 121–128, 175, 178, 220–221, 223, 229
  Ministry of the Environment, 46, 84–86, 89–90, 92
  See also Slovakia

D

Danube River, 1, 2, 8, 138, 143, 146, 147, 148, 167–180, 232
deforestation. *See* forests
Denmark, 47, 99

direct foreign investment, 223
Dörögd Basin, 9, 232–245
double-dividend solution, 99
drinking water. *See* water

E

Ecoglasnost, 2
ecological footprint, 121, 125–126
ecological tax reform, 88, 99–117
education, 8, 9, 19, 31, 48–51, 66, 116, 156, 175, 204, 206, 209–210, 213, 215, 222, 224–225, 228, 232, 236, 243–245
Elbe River, 146–147, 175
emissions, 6, 14, 20–21, 32, 34–36, 38, 40, 46, 49, 53, 86–87, 91–92, 94–96, 100–104, 107, 111–112, 114–116, 122, 143, 157, 232
    trading of, 5–6, 81–97, 103
    carbon dioxide (CO2), 32–34, 40, 42, 84–86, 88–89, 95, 102–105, 109, 111, 116
    sulpher dioxide (SO2), 32, 40, 42, 46, 84, 102–104, 111, 116, 232
    See also pollution
energy, 5–6, 8, 14, 33, 53, 59, 81–85, 87–90, 92, 94–97, 102–104, 107–115, 117, 124, 126–127, 193, 195, 205, 219, 234, 236, 239, 244–245
    renewable, 6, 87–90, 92–94, 236, 239, 244–245
    See also coal
Environment for Europe (EfE). *See* United Nations Economic Commission for Europe (UNECE)
Environmental Partnership for Central and Eastern Europe, 49
environmental protection investment, 100–101
environmental remediation, 47–48, 51
environmental space, 7, 121, 125–126
Estonia, 22–23, 33, 50–51
externalities, 39, 62–63, 73
European Union (EU)
    accession to, 3–5, 7, 45–59, 67, 73, 81, 83, 87, 94, 98, 101, 137, 143, 150, 156–158, 179, 218, 229, 234
    acquis communautaire, 50, 144, 179
    environmental legislation and policy of, 5–7, 46, 49–53, 55, 83–84, 87–88, 94–95, 97, 100–101, 143–145, 148, 156–157, 160, 165, 178–179
    European Environmental Agency, 156
    European Investment Bank, 50
    Financial Instrument for the Environment (LIFE), 50–51
    Instrument for Structural Policies for Pre-Accession (ISPA), 50, 176
    Poland and Hungary Assistance for Economic Restructuring (PHARE), 49–51, 55, 86, 176, 207
    Special Accession Programme for Agriculture and Rural Development (SAPARD), 50, 234–235
    Sixth Environment Action Programme (2002), 51
    Water Framework Directive, 7, 143–149, 156, 178, 180
    Water Law, 148, 156

F

floods, 6–7, 14, 121, 124, 127, 160, 164, 192, 195
forests, 2, 8, 12, 14, 32, 85, 116, 121–122, 126, 151, 155–157, 160, 170, 177, 183, 185–196, 212–213, 235, 237–239, 244
    deforestation, 32–33, 188, 191–192, 238

G

Gabčikovo-Nagymaros Barrage System, 2
Galicia, 8, 183–186, 189, 193. *See also* Carpathians; and Hutsul region
Gross Domestic Product (GDP), 26–30, 36–39, 41, 52, 82–83, 88, 101, 113, 124, 168–171, 177, 203
Gross National Product (GNP), 20, 23–26, 28–29, 33–34
general equilibrium modeling, 100, 103, 116. *See also* computable general equilibrium modeling
Georgia, 22–23, 33
German Democratic Republic (GDR), 2
Germany, 1, 12, 14, 47, 53, 61, 99, 122, 124–126, 146, 167, 169, 172–173, 175, 178

globalization, 5, 9, 71, 217–219, 227
global warming, 13–14, 40
greenhouse gases, 5, 13–14, 81–98, 232
Greenpeace, 2
"green tax reform", 99

H
habitat, 52, 137, 185, 191, 237, 239
heap-leaching, 134
holistic methods of regional geography, 221
Hungary, 1, 3, 7, 9, 14, 20–23, 33, 35, 41, 49–51, 126, 129–135, 137, 139, 147, 169, 173, 178, 205, 220, 232–245
    national agrarian-environmental program (NAKP), 234
Hutsul region, 8, 183–184, 188–191, 193–196, 198. *See also* Galicia; and Carpathians

I
Independent Ecological Center (IEC), 236
industrial districts, 61–64, 67, 74–75
innovation systems, 62, 68, 74
Institute of Geonics, 218
inventory of the state of the environment, 122, 124
Instrument for Structural Policies for Pre-Accession (ISPA), *see* European Union (EU)
Italy 61, 67, 75, 99, 168

K
Kosiv district, 188–189
Kyoto Protocol, 5, 83–84, 87–88, 90, 94, 97

L
labor, 64, 99, 104–106, 109, 113, 115–116, 187, 193–194, 204, 206, 212, 214–215, 223–224
lakes, 14, 121, 138–139, 143, 151–153, 157, 192, 236–238, 242
Latvia, 22–23, 33, 50–51
Leopold, Aldo, 183, 198
liberalization, 4, 20–22, 24–27, 30, 32–34, 37, 40
life expectancy, 20, 30–31
Lithuania, 22–23, 33, 50, 98

M
Macedonia, 22–23, 33
market economy, 1, 8, 20–21, 26–30, 33–35, 38, 49, 53, 61, 62, 64–66, 69–74, 82–83, 87–88, 92, 94, 96, 104–108, 110–111, 114, 124, 131, 194, 203, 208, 215, 218, 222–223, 225, 236
    market-based mechanism (MBM), 4–5, 36
Middle Spiš, 207–209, 214–215
mining, 7, 12, 123, 129–137, 139, 140–42. *See also* coal
Moravia, 9, 123, 217–231

N
Nadvirna district, 188–189
national agrarian-environmental program (NAKP). *See* Hungary
NGOs, 33, 47–49, 51–52, 54, 125–126, 130–131, 138–139, 197, 214, 235–236, 242
nuclear power, 2, 12, 52, 94, 98, 126, 154–155, 159
    Temelín nuclear power station, 2

O
Organisation for Economic Co-operation and Development (OECD), 4, 30, 62–63, 74, 84, 100, 111, 174
overpopulation, 31

P
Poland, 1, 5–6, 8, 20–23, 33, 35, 41–42, 46, 49–50, 52–53, 84, 99–117, 126, 150–166, 168, 187, 205
    See also Silesia (Polish)
Poland and Hungary Assistance for Economic Restructuring (PHARE). *See* European Union (EU)
pollution, 2, 4–6, 12, 20, 34–36, 45–47, 50, 53, 89, 99, 124, 232
    Air, 33, 38–39, 43, 81–84, 87–88, 102–103, 107, 109, 111–112, 115, 124, 232
    Water, 8, 122, 124, 129, 131, 134, 147, 151, 153–155, 160–161, 167–169, 171, 173–174, 176–179, 194, 221, 233
    See also emissions

privatization, 20–27, 29–31, 34–35, 37–38, 40, 173, 197, 222–223
Prut River, 193–194

R

Regional Environmental Center (REC), 49, 51
Rhine River, 137, 146–147, 167
Roma, 8–9, 203–216
Romania, 1, 2, 4, 7, 20, 22–23, 33, 46, 50–51, 129–134, 137–140, 147, 169, 173, 179. *See also* Aurul, S.A.; Baia Mare; Baia Borsa; and Rosia Montana
Rosia Montana, 130, 137
Russia, 5, 12, 14, 20–23, 29, 33, 35–36

S

Special Accession Programme for Agriculture and Rural Development (SAPARD). *See* European Union (EU)
Serbia, 133, 138–139
Sewage, 2, 52, 122, 144, 168, 170, 172, 173–175, 177–180, 205, 226, 233, 244–245
Silesia (Polish), 2
Sixth Environment Action Programme (2002). *See* European Union (EU)
Slovakia, 3, 8, 20–23, 33, 41, 50–51, 82, 84, 94, 123–124, 168–169, 172–173, 175–178, 203–216, 222, 224
Slovenia, 22–23, 33, 50–51, 83, 98, 147, 169, 173
small-town sector, 9, 217–229
Stockholm Declaration on the Human Environment, 140
strategy for sustainable development, 122, 124–127
"structural adjustment funds," 52
Switzerland, 136–137, 140, 168

T

taxation, 6, 49, 53, 88, 97, 99–117, 134, 137, 206, 215

Temelín nuclear power station. *See* nuclear power
Tisza/Tisa River, 2, 7, 129–141
tourism, 125, 170, 193, 221, 225, 228, 236, 244–245
traffic, 226, 242
Turkey, 50, 238

U

Ukraine, 2, 8, 12, 22–23, 33, 150–156, 159–164, 169, 184, 186, 196–197, 224
United Nations Development Program (UNDP), 169, 171, 174, 176, 207
United Nations Economic Commission for Europe (UNECE), 7–8, 139, 150–165
    Aarhus Convention, 51
    Environment for Europe (EfE), 48

V

Verespatak. *See* Rosia Montana
Vistula River, 2, 152

W

water
    drinking water, 52, 133, 137, 144, 147, 153, 166n33, 170–171, 205, 213
    Water Framework Directive, *see* European Union (EU)
    Water Law, *see* European Union (EU)
    management of, 7, 101, 143–148
    wastewater, 8, 46, 50, 52, 130, 132, 136, 143, 147, 153, 167–180, 233
welfare, 26, 30, 99, 111–113, 205, 209–210, 213, 224
Western Bug River, 7–8, 150–162
women, 4, 13, 210–211, 213
World Bank, 4, 33, 38, 48, 176, 205

Y

YOUR SPIŠ, 208–215